Revolutionary Atmosphere

Revolutionary Atmosphere

The Story of the Altitude Wind Tunnel and
the Space Power Chambers

By Robert S. Arrighi

National Aeronautics and Space Administration

NASA History Division
Office of Communications
NASA Headquarters
Washington, DC 20546

SP–2010–4319
April 2010

Library of Congress Cataloging-in-Publication Data

Arrighi, Robert S., 1969-
Revolutionary atmosphere: the story of the Altitude Wind Tunnel and the Space
Power Chambers / by Robert S. Arrighi.
  p. cm. - - (NASA SP-2010-4319)
"April 2010."
Includes bibliographical references.
1. Altitude Wind Tunnel (Laboratory)—History. 2. Space Power Chambers
(Laboratory)—History. 3. Aeronautics—Research—United States—History.
4. Space vehicles—United States—Testing—History. I. Title.

TL567.W5A77 2010
629.46'8—dc22

               2010018841

For sale by the Superintendent of Documents, U.S. Government Printing Office
Internet: bookstore.gpo.gov Phone: toll free (866) 512–1800; DC area (202) 512–1800
Fax: (202) 512–2104 Mail: Stop IDCC, Washington, DC 20401–0001

Contents

Chapter 4

Chapter 5

Chapter 6

Chapter 7

Chapter 8

Chapter 9

Chapter 10

Chapter 11

Chapter 12

Chapter 13

The Altitude Wind Tunnel's (AWT) steel shell loomed, almost threateningly, over the National Aeronautics and Space Administration (NASA) Glenn Research Center's main campus in Cleveland, Ohio, for over 60 years. The facility had been inactive since 1975, but the hulking tunnel sat in a conspicuous location between the visitors' center and the Icing Research Tunnel and was seen daily by hundreds, if not thousands, of people. The tunnel slowly ebbed from NASA Glenn's collective consciousness. Inside the steel shell, significant contributions had been made in the advancement of the jet engine and the high-profile Project Mercury and Centaur Program. Yet, the AWT had remained a mystery to most current employees and the public. Not only did the rusting giant have an obscure past, few even knew its name. This book, the accompanying Web site *(http://awt.grc.nasa.gov)*, and other documentation have been created to resurrect the esteemed reputation of this once-vital and historically significant facility.

The AWT's unrivaled capability to test full-scale engines in simulated altitude conditions advanced the development of the jet engine considerably during its formative period in the 1940s and in its maturity in the 1950s. The AWT was the nation's first wind tunnel built specifically to study the operation of engines. Its ability to consistently re-create flight conditions allowed researchers to systematically study engine behavior and perfect innovations such as the afterburner and the variable-area nozzle.

Between 1959 and 1963 the AWT was slowly transformed into two large test chambers. The tunnel's simulated high-altitude conditions allowed NASA to cancel costly and time-consuming flight testing for Project Mercury. Afterward the tunnel was converted into one of the nation's first large

Image 1: For over 60 years, the massive AWT was a central fixture at what is today the NASA Glenn Research Center. Despite its prominent location, few current employees were aware of its storied past or major contributions to aeronautics and space. (NASA C–2005–01674)

vacuum chambers and renamed the "Space Power Chambers" (SPC). It was used to quickly remedy a number of problems for the Centaur second-stage rocket. The SPC tests allowed the Centaur to sustain its tight schedule for the Surveyor and later orbiting observatory missions. Use of the facility tapered off in the 1970s, and an effort to resurrect the wind tunnel failed in the early 1980s. After years of neglect, the tunnel was demolished in late 2008 and spring 2009.

NASA's historical publications tend to focus on center histories, specific programs (particularly the human space program), or chronologies and other reference materials. The NASA Glenn Research Center has carved out a niche by writing the histories of several of its research facilities. The Icing Research Tunnel, Rocket Engine Test Facility, and Plum Brook Reactor have all been documented in recent years. These three books appear to be the only single-facility studies in NASA's historical collection.

Certainly the center and program histories describe test facilities, but they are mostly portrayed as research tools contributing to a larger complex or project. The histories of facilities, therefore, provide a unique perspective. In some cases, the facilities have made long-term contributions to a specific program or field, such as the Icing Research Tunnel and the field of icing research. Other facilities, however, can serve as a useful lens through which to view the progression of research, technology, and the larger laboratory over a long period of time. The AWT, with its 30-year operational career and its contributions to the turbojet and space revolutions, falls into this latter category.

The story is important for several reasons. First, it demonstrates that the ability to adapt to technological changes is vital for large test facilities. Long-term investment in test facilities differentiated the National Advisory Committee for Aeronautics (NACA) from manufacturers, the military, and its counterparts in other countries. The constant evolution of the aerospace field has made it important for facilities to be flexible and strong enough to support upgrades and repurposing.

After less than two years of construction, the AWT came online in February 1944. The tunnel itself was 263 feet long on the north and south legs, and 121 feet long on the east and west legs. The larger west end of the tunnel was 51 feet in diameter throughout. The east side of the tunnel was 31 feet in diameter at the southeast corner and 27 feet in diameter at the northeast. The test section was 20 feet in diameter, but with the introduction of turbojets, the air was ducted directly to the inlet. The complex originally included five support buildings. A water pump house, additional exhausters, and an exhaust gas cooler were added in 1951.

The tunnel shell retained its basic dimensions during the conversion into large test chambers in 1961, although its internal components were removed and bulkheads were added to section off the chambers. The addition of a 22.5-foot-diameter extension and dome was the largest visible change. This added about 14 feet to the height of the vacuum chamber. A vacuum pump house was added, and use of the water pump house and exhaust cooler ceased. Throughout its lifespan, the tunnel grew, adjusted, morphed, attempted to return to its original function, and was eventually demolished. By constantly changing, the tunnel was able to remain a vital tool for over 30 years.

Second, the history of the AWT provides a look at the growth of the Lewis laboratory[1] and the dramatic progression of aerospace technology during its most fertile years. With World War II approaching, the NACA realized it needed a laboratory and wind tunnel to study large piston engines. Although not operational until early 1944, the AWT did remedy a serious cooling problem for the B-29 bomber's engines. More importantly, though, from its very first runs, the AWT was used to study and improve the newly developed turbojet engines. The tunnel's altitude capabilities were ideally suited to studying the problem of flameout in early jet engines. At the military's request, nearly every type of jet engine during the 1940s and early 1950s in the United States underwent evaluation in the AWT. By the 1950s the performance of turbojets was increasing dramatically.

Although the AWT underwent several modernizations, its capabilities were eventually superseded in the late 1950s by the laboratory's newer supersonic tunnels and altitude test cells. The internal components were removed from a large section of the AWT in 1959. This area was used as an altitude chamber for several Project Mercury test programs, including the attitude control system for the Big Joe launch. In 1960 all seven of the Mercury astronauts came to the AWT to train in the Multi-Axis Space Test Inertia Facility (MASTIF).

Afterward the tunnel was permanently converted into two test chambers. The new SPC contained a large vacuum tank used to simulate the 100-mile-altitude atmosphere of space. When it became operational in September 1963, the new 70,000-cubic-foot vacuum tank was rivaled in size only by the Mark I tank at the Arnold Engineering Development Center.* Although the NASA Lewis Research Center created several smaller vacuum tanks to study ion engines in the Electric Propulsion Laboratory, the SPC contained Lewis's primary altitude chambers throughout the 1960s.

The two chambers in the SPC were used to remedy several problems with the Centaur rocket during the lead-up to its missions to soft-land the *Surveyor* spacecraft on the Moon. The Centaur-Surveyor missions were a key precursor to the Apollo lunar landings. Later, as the Centaur payloads grew

---

*The Cleveland laboratory began operation in 1942 as the NACA Aircraft Engine Research Laboratory. In 1947 it was renamed the NACA Flight Propulsion Laboratory to reflect the expansion of the research. Following the death of the NACA's Director of Aeronautics, George Lewis, the name was changed to the NACA Lewis Flight Propulsion Laboratory in September 1948. On 1 October 1958, the lab was incorporated in the new NASA space agency, and it was renamed the NASA Lewis Research Center. Following John Glenn's flight on the space shuttle, the center name was changed again on 1 March 1999 to the NASA Glenn Research Center.

in size, the SPC was used to qualify new shroud configurations for the first space observatory satellites.

A third reason why the story of the AWT and its later incarnation as the SPC is important is that it offers a unique view of a facility and a laboratory from both sides of the October 1958 Space Act. There are several areas of differences between NASA and its predecessor, the NACA, including the organizational structure, the use of contractors, and the size of the budget. The differences most often cited are research versus development and the drift of personnel from applied engineering to management. The NACA's emphasis on research resulted in the construction of large facilities, particularly wind tunnels during the committee's 20 years of greatest achievement, 1935 to 1955. The tunnels were used almost entirely for research activities. With the advent of the space program, equally impressive facilities were built, but they were used primarily to test or verify items already in development. This is illustrated by the AWT, which was primarily used for research, and the SPC, which focused on development.

Image 2: The mechanics and technicians associated with the AWT were a crucial element of the facility's success. As technology improved and the Agency's priorities changed, the staff improved their skills and knowledge. This was particularly important during the transition from piston engines to turbojets and from aeronautics to space. (NASA C–1949–23127)

This book emphasizes the importance of the technical staff and their skills to the AWT's success. Operating the facility required the integrated work of a complementary group of individuals. From mechanics and technicians to test engineers and scientists, the group at the Cleveland lab was exceptional. The mechanics and technicians, many of whom came through the Apprentice Program, were often as responsible for the success of a test as the researchers. Together the group solved problems affecting engines and developed new methods of improving performance. As the focus shifted to the space program, the staff had to adjust to the precision required for space hardware. The organization was often pushed to its limits, particularly during World War II and the first years of the Centaur Program in the early 1960s. Many of the chapters include a section highlighting the personal stories of the AWT staff.

The AWT story demonstrates the importance of leadership to the success of a facility and a research laboratory. This is exemplified by Abe Silverstein, one of the true visionaries of both the NACA and NASA. He was a hands-on manager who was personally involved with the work of individuals and specific projects. His technical expertise not only led to resolutions of difficult problems but gained him the respect of staff and the Agency's leadership. He had the ability to grasp the long view and to identify new fields that would lead to future advances. The AWT's missions were often directly impacted by Silverstein's decisions.

Silverstein served as the first head of the Lewis laboratory's AWT Section and then as the chief of the Wind Tunnel and Flight Division. He was responsible for testing the first U.S. jet engines in the AWT and for the construction of the lab's first supersonic wind tunnels. In 1949 Silverstein became director of all research at the lab, and in 1952 he became Associate Director of Lewis. In these roles, he made sure that nontraditional areas such as liquid-hydrogen fuel, rocket engines, electric propulsion, and nuclear propulsion were explored. He continued pushing the aeronautics work as well, particularly on transonic and supersonic compressors for axial-flow turbojet engines.

Silverstein was transferred to Washington in early 1958 to help Hugh Dryden create the nation's new space agency. Afterward Silverstein played a key role in arranging the manned space program, planned numerous satellite and interplanetary missions, and led a committee that selected a liquid-hydrogen upper stage for the Saturn rocket. Silverstein returned to Cleveland in 1961 to serve as Director of the NASA Lewis Research Center. He was responsible for the transfer of the Centaur rocket to Lewis in 1962, and personally oversaw the project until it became successful. Silverstein retired in 1969 as the

Image 3: Abe Silverstein prepares for a talk on the AWT's research. Silverstein was the original manager of the AWT. As he moved up through the ranks of management, he always remained close to the research work being done at the Cleveland lab. His decisions affected the AWT's research for over 25 years. (NASA C-1946-14511)

Apollo Program reached its zenith. His career was intertwined with the story of the AWT and is highlighted throughout.

The AWT's accomplishments would not be possible without this trinity of excellence—the large ever-evolving test facility, the expertise of the researchers and mechanics, and the direction provided by Silverstein. The AWT story also provides an insightful look at the advancement of turbojet engines, the early days of the space program, the development of the Centaur second-stage rocket, and the shift from the NACA to NASA.

In the 1970s the facility was again superseded, this time by the world's largest vacuum chamber, the Space Power Facility at NASA Glenn's Plum Brook Station. The SPC remained largely vacant during the late 1970s. A major effort was undertaken in the early 1980s to determine the feasibility of converting the facility back into a wind tunnel for a new generation of testing. Congress rejected the proposal in 1985, and the facility remained idle. During the ensuing 30 years of dormancy, the facility began suffering from the neglect. In 2004 NASA initiated a large effort to identify and demolish underutilized facilities. Because of the AWT's lack of mission, high maintenance costs, and environmental hazards, NASA Glenn began planning for the removal of the facility. Although surveyed in 1996, no determination was made on its eligibility for the National Register of Historic Places. Nonetheless, Glenn was aware of its significance and began documenting it as though it were eligible for listing before demolition began in 2008. It is hoped that this book will preserve the AWT's considerable legacy.

Endnotes for Preface

1. R.T. Hollingsworth, *A Survey of Large Space Chambers* (Washington, DC: NASA TN D–1673, 1963), p. 12.

Acknowledgments

This project was born out of an arrangement coordinated by former NASA Glenn History Officer, Kevin Coleman, and former Glenn Historic Preservation Officer, Joseph Morris, to document the AWT prior to its demolition. Kevin and Joe had previously worked together on the historic mitigation work for Glenn's Rocket Engine Test Facility. That project hired outside contractors to perform much of the work. In an effort to reduce the cost for the AWT mitigation and utilize the skills of Glenn's Logistics and Technical Information Division (LTID), Kevin Coleman and Wyle Media Services Manager on the TIALS Contract Ralph Bevilacqua put together a proposal to handle all the mitigation work in-house. An agreement between LTID and the Facilities Division was finalized in May 2005.

It was just another in a long string of entrepreneurial enterprises that Kevin had put together during his nine years serving as Glenn's History Officer. Kevin retired in September 2008 and is missed as a friend, confidant, and mentor. Not long after the project began, Leslie Main took over the duties of the center's Historic Preservation Officer. Les has been extremely helpful on all levels. He has provided much needed technical insight, contacts with construction and facility managers, and a common sense approach to the mitigation work. I am currently working with Les on a number of smaller projects relating to other Glenn facilities.

At Headquarters, Jane Odom, John Hargenrader, and Colin Fries were very helpful during my research in the NASA History Division. Glenn Bugos at the NASA Ames Research Center graciously arranged to interview Walter Vincenti for me and reviewed the first two chapters.

Wyle Information Services employees on the TIALS contract assisted with many different aspects of the work. Suzanne Kelly and Debbie Demaline were extremely helpful in identifying and retrieving documents and negatives stored in the Glenn Records Management System. Jan Dick located several hard-to-find technical reports. Mark Grills and Michelle Murphy performed the daunting task of scanning hundreds of negatives, and Quentin Schwinn and Bridget Caswell took excellent photographs at the facility prior to and during its demolition. Nancy O'Bryan and Lorraine Feher helped with copyediting and proofreading, and Kelly Shankland prepared the cover design and the layout of the document.

Mark Bowles, Galen Wilson, Tina Norwood, and anonymous peer reviewers took the time to thoroughly review the manuscript. Their comments, suggestions, and questions improved the material significantly. Mark, Galen, and Tina not only read the text closely, but also discussed the material with me at length. I really appreciate their extra efforts. Thanks also go to Steve Garber at the NASA History Division and Glenn History Officer Anne Power for coordinating the book reviews and publication.

Several retirees contributed to the book, but the assistance of Bill Harrison, Frank Holt, and Larry Ross was exceptional. They generously helped tell the story, located other retirees and resources, reviewed chapters, and identified photographs. More importantly they gave me a real sense of the family atmosphere at the lab and the underappreciated role of the mechanics and technicians.

I also want to thank Virginia Dawson and Mark Bowles, who provided me with my first work as an archivist almost a decade ago and who later introduced me to NASA. Their professionalism and zeal have inspired me to pursue a career in the history of technology and of NASA in particular.

Finally, I want to thank my wife Sarah, who has supported me throughout this project and who has been, for the most part, patient with me as I worked on the manuscript in the evenings, on weekends, and while away on vacation. Thanks go to my family, Bob Sr., Gerri, Michele, Dan, and Autumn for everything they have done for me over the years.

Image 4: The Hindenburg airship crossing the Atlantic on 4 April 1936. George Lewis would cross on the airship several months later on his epic tour of German research laboratories. (National Archives LC–USZ62–70162)

Chapter 1

Premonition | The Need for an Engine Tunnel (1936–1940)

With only the cabin windows, engine nacelles, and running lights illuminated, the new German luxury airship hung surreally in the darkened sky as Dr. George Lewis arrived at the field on the night of Monday, 21 September 1936. Dr. Lewis and the other passengers filed up the long gangway into the hovering ship. Upon reaching the top of the stairs leading to the main deck, Lewis came face to face with a bust of Field Marshal Paul von Hindenburg.[2] Shortly afterward, the *Hindenburg* left its mooring and silently carried Lewis away from Lakehurst, New Jersey, on a journey that would change the way the United States approached aeronautical research.

The 54-year-old aviation veteran was impressed by the massive airship's calm and steadiness, even later when traversing the wake of a passing hurricane. As Director of Aeronautical Research for the National Advisory Committee for Aeronautics, Lewis was responsible for coordinating the nation's aviation research. The *Hindenburg's* silent comfort during the 55-hour passage may have only increased Lewis's foreboding thoughts on what lay ahead of him across the ocean.

Lewis's journey to Germany and Russia was spurred by an invitation from the Deutsche Zeppelin-Reederei to tour the German aeronautical facilities. Despite the dominance of U.S. air manufacturers, Lewis was aware that the Europeans were ahead in the development of high-powered engines.[3] A letter from the NACA's European liaison, John Jay Ide, had already piqued Lewis's interest with its description of a large pressurized wind tunnel at Göttingen, and visiting Russian researchers had claimed that their aeronautical research institute had 3,500 employees.[4] The NACA consisted of a single research laboratory, Langley Aeronautical Laboratory, with only 350 employees.

Both the economy and aircraft industry had boomed in the United States following World War I while Europe suffered. By the time of Lewis's trip in 1936, Boeing and Douglas possessed the premier airlines in the world. During this period Europeans were exploring ways to fly higher and faster, while Americans were easing up on their research and development. As a result, the United States had greater quantities of aircraft but would find itself lagging behind in regards to propulsion advances such as liquid-cooled and turbojet engines.[5]

"I Have Seen Nothing Like Them in America"

German Chancellor Adolf Hitler had created an air ministry, the Deutsche Versuchsanstalt für Luftfahrt, and supplied it with ample funding for extensive research and development. Lewis's personal guide in Berlin was

Image 5: Dr. George Lewis managed the aeronautical research conducted at the NACA for over 20 years. His most important accomplishment, how-ever, may have been an investigative tour of research facilities in Germany in the fall of 1936. A second tour in 1939 included Germany, Britain, and France. The visits resulted in the NACA's physical expansion and the broad-ening of the scope of its research. (NASA EL-1997-00143)

Dr. Adolf Baeumker, the head of all German aeronautical research and development projects. When Baeumker told Lewis, "the ordinary military man is only interested in getting a large number of airplanes and not in aeronautical development and research," he was praising Air Minister Hermann Göring, but he could just as well have been referring to the field of aviation in their respective countries.[6]

Baeumker explained that Germany had developed a long-term aeronautical research plan that involved three major research laboratories. In what would become a model for the NACA, the Deutsche Versuchsanstalt für Luftfahrt had expanded its primary laboratory near Berlin and created two new research labs—one for aerodynamics and the other for engine research. Lewis estimated the total employee contingent in the German air ministry to be 1,600 to 2,000 and growing. Lewis remarked (about the German laboratories), "I have seen nothing like them in America." Lewis concluded, "If the United States is to hold its present position in the field of aeronautical research, it will be necessary to increase the personnel at Langley Field [Langley Aeronautical Laboratory]."[7]

Lewis was also shown a 22- by 17-foot pressure tunnel at Göttingen University. Göttingen had been the site of the first modern wind tunnel and had helped to forge the basis of the internally supported wing span. Lewis referred to the facility, which was capable of creating pressure levels three times that found at sea level while maintaining a useful Reynolds number, as the "first wind tunnel of this type to be constructed in the world."[8] Lewis later confided that he was likely the first foreigner to ever visit the Göttingen facility.[9]

Evolution of the Wind Tunnel

The concept of trying to re-create flight in Earth-bound facilities had developed during the 19th century in Europe. The Wright Brothers constructed a small wind tunnel in 1901 that yielded significant lift information for their early flights. However, it was Albert Zahm's tunnel at Catholic University, also constructed in 1901, which became the first significant U.S. tunnel. The 50-foot-long tunnel with its 6- by 6-foot test section dwarfed any of its contemporaries.[10] Zahm's tunnel and those following it benefited from replacing steam engines with more efficient electric-powered engines, which resulted in greater wind speeds at a lower cost.[11] Though Zahm's tunnel was hampered by occasional electrical variations and the effect of external weather conditions, the airflow controls and instrumentation would be used by other tunnels for years.[12]

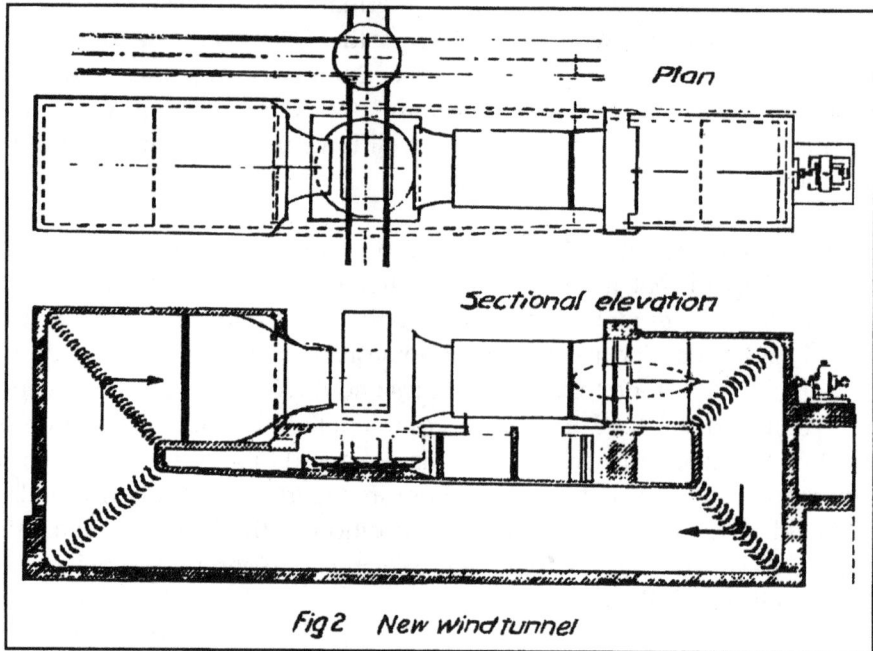

Image 6: *Ludwig Prandtl's 1917 wind tunnel at Göttingen University in Germany. Prandtl's closed-loop tunnel revolutionized wind tunnel design. It included a throat section, turning vanes, and an airflow straightener. Studies in the Göttingen tunnel included wing profile examinations, full-scale propeller tests, and studies of the boundary layer on a rotating cylinder.[13] (1920) (NACA TN No. 66, Fig. 2).*

Despite these early advances, the field of aeronautics in the United States was led primarily by nonprofessionals, whereas in Europe many scientists and engineers were pushed into the field by their governments. In addition, U.S. military leaders did not appreciate the importance of aircraft. The lack of research and government interest led to a deficiency of aircraft in the nation that pioneered aviation. An NACA report states, "When World War I erupted in 1914, it was reported that France had 1400 airplanes, Germany 1000, Russia 800, Great Britain 400, and the United States 23!"[14]

Russia, France, and Great Britain all constructed substantial wind tunnels around the turn of the century. The most influential European facility was Ludwig Prandtl's tunnel built at the University of Göttingen. With funding from the German Society for Airship Study, Prandtl had his initial tunnel operating by 1909. It was the first closed-loop tunnel and the first to use turning vanes or "deflectors" to guide the airflow around the corners. From the beginning, this facility was seen as a stepping stone to a bigger, more complex tunnel. The design of this new tunnel had begun in 1911. After being delayed by the war, the new facility became operational in the spring of 1917.[15]

Prandtl's new tunnel expanded upon the innovations of the first. Its 120-mile-per-hour (mph) speeds made it the most powerful wind tunnel of the period. The tunnel's diameter was expanded to 20 square meters upstream and then narrowed sharply to 4 square meters just before the test section to increase the airspeed. A honeycomb screen was installed across the width of the tunnel to straighten the airflow without slowing its speed. The basic components of Prandtl's tunnels were standard on almost all later tunnels, including the AWT.[16]

The NACA Starts Its Tunnel Collection

Although its formation in 1915 was too late to have much of an influence on World War I, the NACA's staff and mission were substantially expanded after the war. In 1917 the NACA established its own research lab at Langley Field in Hampton, Virginia. Unlike those found in industry, the NACA Langley Aeronautical Laboratory's facilities were to be used primarily to study new aircraft designs and tackle anticipated future aeronautical problems for the civil aviation industry. One of the principal reasons for creating the lab was to build wind tunnels. The first was a small insignificant tunnel that became operational in 1921. Coincidentally this tunnel, the Atmospheric Wind Tunnel, was also referred to as the "AWT."[17]

Langley's next tunnel, the Variable Density Tunnel, however, was a major advancement. It was proposed in 1921 by Dr. Max Munk, who had worked under Prandtl at Göttingen before emigrating to the United States. The Variable Density Tunnel was the first tunnel to forgo normal airflow for highly compressed air. It included a large steel tank in which the atmosphere could be pressurized and a wooden test section that negated the Reynolds number differential between tunnel tests and actual flight conditions. The tunnel, which became operational in 1923, was a highly useful tool and temporarily closed the gap between U.S. and European research.[18]

Langley began putting an entire collection of increasingly complex wind tunnels into operation. The next was the Propeller Research Tunnel in 1927, followed by the Vertical Spin Tunnel and Atmospheric Wind Tunnel in 1930, the Full-Scale Tunnel in 1931, the 8-Foot High Speed Tunnel in 1936, and the 19-Foot Pressure Tunnel in 1939. The 8-Foot High Speed Tunnel, which could simulate pressure altitudes of 12,000 feet and speeds of 500 mph, and the 19-Foot Pressure Tunnel, which combined a large test section with 250-mph speeds, were significant steps forward in flight simulation.[19]

Image 7: Elton Miller, successor to Max Munk as Langley's chief of aerodynamics, stands in the exit cone of the Propeller Research Tunnel to view the Sperry M-1 Messenger during the tunnel's first runs. The Propeller Research Tunnel at Langley was used to clean up drag caused by aircraft engines but not to study the performance of those engines. The tests were run in atmospheric conditions and at relatively slow speeds. (1927) (NASA EL–01892)

Exile of the Engine

Early research at Langley dealt almost exclusively with aerodynamics, even in its propeller tunnel. The Propeller Research Tunnel could fit an entire aircraft with an operating engine in its 20-foot-diameter open test section and could test it at airspeeds up to 110 mph. The tunnel made considerable contributions to the aerodynamics of propellers and engine nacelles, including the "NACA Cowl," but little to the actual operation of the engines.[20] Langley's Powerplants Division, led by Carlton Kemper, had only a handful of people until 1938, when it increased to 12.[21] The group dealt with the fundamentals of engine power, efficiency, and fuel consumption, but it did not delve into specific problems associated with these elements.[22]

The NACA's lack of interest in propulsion seems to have stemmed from an early meeting that it held in 1916 with automobile and aviation engine manufacturers. The group agreed that with military funding for research, these manufacturers would possess the needed assets and skills to develop aircraft engines. Initially the NACA did not have facilities to test aircraft engines, so Samuel Stratton, who headed the National Bureau of Standards and served as an original NACA committee member, took on the testing for the manufacturers at the National Bureau of Standards laboratories. The NACA seemed content with this arrangement even after Langley was established. As late as 1937, an original NACA member, Dr. Joseph Ames, stated, "The [engine] problem is primarily and almost solely one of development, which can best be attacked by the aircraft engine industry under experimental contracts with the War and Navy Departments."[23]

In order to stretch limited propulsion funding and manpower, the Power-plants Division primarily studied single cylinders and then extrapolated the test data to full-scale engines. Abe Silverstein, who was working with the aerodynamicists in the Full-Scale Tunnel at Langley, discovered that each cylinder produced a unique temperature during engine operation. This phenomenon could not be studied using single cylinders. Individual engine components could operate well in isolation only to fail when integrated with the larger system. Langley was not equipped to test large full-scale engines, and Kemper continued to put the engine onus on the manufacturers.[24]

The coordinated national research program in Germany pursued engine technology as well as aerodynamics. On his 1936 tour, Lewis was also shown a Deutsche Versuchsanstalt für Luftfahrt facility that was able to test water- and air-cooled engines under simulated altitude conditions. The intake air and exhaust were kept at pressure and temperature levels that corresponded to altitude conditions. The extensive instrumentation and infrastructure led Lewis to wonder "whether one is in an engine-testing laboratory or a small edition of the Licke observatory."[25]

Construction of a larger facility to test engines would begin at Göttingen during World War II. It was a wind tunnel designed to study operating full-scale jet engine systems in simulated altitude conditions at speeds up to 290 mph. Its refrigeration system would produce temperatures of –64°F, allowing the tunnel to also be used for icing studies.[26] Although construction would only be 80 percent complete because of severe German defeats during the war, the tunnel would have been remarkably similar to the AWT in Cleveland.

Engine testing was taking place in the United States at this time, but it was scattered and uneven. Private companies could not afford to build large test facilities. The National Bureau of Standards, the U.S. Naval Aircraft Factory, and the U.S. Army Air Corps at McCook Field had successfully designed tanks to test small engines that could simulate the temperatures and pressures associated with altitude, but they were of limited size and could not incorporate the benefits of a wind tunnel.[27] On the other hand, the 7- by 10-foot Wright Brothers Wind Tunnel at the Massachusetts Institute of Technology and the 19-Foot Pressure Tunnel at NASA Langley, both completed in 1939, could simulate altitudes in a wind tunnel setting but were not capable of running aircraft engines. The AWT and the unfinished Göttingen tunnel were intended to solve both sides of the equation.

Peace-Time War

George Lewis was not the only NACA member visiting Germany in the 1930s. Charles Lindbergh was living abroad to escape the publicity surrounding the kidnapping and murder of his son. From this vantage point, he was able to provide unique insight into the European aeronautical industry. In 1936 Lindbergh made the first of five trips to inspect German aeronautics facilities. He regularly wrote to Dr. Ames with his findings.

In a letter dated 28 November 1938, Lindbergh wrote that European military aircraft were willing to trade range for ever-increasing speed and altitude. He felt that the United States should concentrate its efforts on improving its aircraft designs instead of building additional aircraft. He concluded, "It is more necessary than ever before to give full scope to research if we are to regain the leadership we have lost in some of the fields of military aviation."[28] Ames responded the following week that General Oscar Westover, then Chief of the U.S. Army Air Corps, had initiated the NACA's new Future Research Facilities Special Committee to investigate the possibility of starting a new NACA laboratory. Ames, added, "There is a new atmosphere in Washington. It has been likened to a state of 'peace-time war.'"[29]

In response to General Westover's special committee report and the impending war, the military finally turned to the NACA for assistance. Congress approved funding for the expansion of the NACA in 1939 but felt that a new laboratory was needed to decongest the Langley research community and disperse the facilities to prevent a catastrophic enemy attack.[30] Initially the NACA sought a single additional facility in which to investigate high-speed flight. Lindbergh returned to the United States to lead a special committee to research sites for the new lab. The group selected Moffett Field in Sunnyvale, California, and construction of the NACA Ames Aeronautical Laboratory began on 20 December 1939.[31]

The Lindbergh committee report also stated, "There is a serious lack of engine research facilities in the United States, and that it is of the utmost importance for the development of aviation in general, and for our defense program in particular, to take immediate steps to remedy this deficiency."[32] It was evident that the NACA also required a laboratory to specifically study aircraft engines.

On 19 October 1939 the NACA endorsed the new research facility for aircraft propulsion systems. George Mead, former vice president of both Pratt & Whitney and United Aircraft Corporation, was selected to lead a Special Committee on New Engine Research Facilities. The team included George Lewis, Carlton Kemper, the U.S. Army's Major Edwin Page, and others. A committee report issued on 23 January 1940 called for a $10-million laboratory that would include a test stand for engines, a fuels and lubricants facility, and—after some debate—a wind tunnel for engines.[33,34]

Image 8: Left to right: Charles Lindbergh, U.S. Air Force Vice Admiral Arthur Cook, U.S. Navy Charles Abbot, and Dr. Joseph Ames. From Europe, Lindbergh wrote to fellow NACA member, Dr. Ames, "I believe we should accept the fact that Germany will continue to be the leading country in Europe in aviation. She will be the leading country in the world if we do not increase our own rate of development. Even now Germany is far ahead of us in military aviation." Lindbergh claimed that Germany's ever-growing research facilities were so new that it would be years before their impact could be determined.[35] (NASA GPN–2002–000024)

Image 9: General Henry Arnold greets fellow NACA committee member Dr. George Lewis. Arnold served on the main committee from 1938 to 1944 and was a strong advocate for a new engine research facility. Arnold believed in continual research and development. Some of his most influential advisors were civilians such as Theodore von Kármán, Robert Millikan, and Charles Lindbergh.[36] (NASA C-1944-07495)

NACA Executive Secretary John Victory remembered seeing a newspaper headline stating "German Army Enters Paris" while sitting in the Senate gallery during the June 1940 appropriation hearings. Although the Bureau of the Budget lowered the requested amount to $8.4 million, the figure was eventually increased to $13.3 million to accommodate a larger role for what was to become the AWT.[37]

Engine Research Tunnel

During the Alessandro Volta Foundation's 1935 conference on High Velocities in Aviation, Jakob Ackeret predicted the use of the gas turbine for high-speed, high-altitude aircraft propulsion. He also stated that new wind tunnels capable of simulating altitude were important since high-speed flights would likely take place at high altitudes. Ackeret envisioned a tunnel in which a vacuum or pressures up to 10 atmospheres could be maintained.[38]

Although the NACA participated in the conference, it was not until the impending war in 1939 that the importance of increasing altitudes for U.S. aircraft was thrust to the forefront of the NACA's consciousness. Improved antiaircraft weaponry would force bombers and fighters to fly higher in order to be effective. Air battles in World War I were generally fought at 100 mph and at an altitude around 10,000 feet. By early 1941, newspapers were reporting that World War II aircraft were already flying 400 mph at 20,000 feet. It was anticipated at the time that aircraft would be exceeding 500 mph and 30,000 feet in the near future.[39]

Image 10: A full-scale aircraft is lifted into the AWT's test section. Although testing full-size versions reduced the cost, time, and effort of developing engines, it also required a substantial investment in large facilities. Many of the NACA's achievements resulted from its willingness to make those investments. (NASA C-1944-06305)

Image 11: Early drawing of the new AWT, which was originally referred to as the "Engine Research Tunnel." It was the nation's first wind tunnel built to study engines under simulated flight conditions. (NASA C–1944–05308).

George Mead's Special Committee on New Engine Research Facilities rapidly began developing ideas and designs for test facilities at the new engine lab. Rudolph Gagg, a design consultant from Wright Aeronautical, said at the time, "This aviation research laboratory is being designed for the engines just around the corner. The purpose of the laboratory is to look forward to the engines of the future."[40] Although this would prove not to be true, the facilities would be robust enough to adapt to the new engines that would be around the corner.

The ability to test power, speed, drag, vibration, and cooling on complete engine systems would provide a faster transition from design to flight testing while avoiding additional time-consuming flight tests with risky and unproven engines. Performance in the harsh temperature and pressure conditions of these higher altitudes was vital.[41] The design and construction of the Engine Research Tunnel would prove to be a monumental task. It would combine elements from NACA Langley's Propeller Research Tunnel, 8-Foot High Speed Tunnel, and 19-Foot Pressure Tunnel with a massive refrigeration system and the ability to run large engines.

Not everyone agreed with the concept. Just weeks after the engine lab was approved, Mead wrote George Lewis, "I very much doubt whether there is a real need for the altitude chamber. In my opinion, a much more useful piece of equipment would be a full-scale tunnel in which we could test engines up to 4000-horsepower at sea-level pressure, the purpose being to check functional characteristics of multicylinder engines and their installations at speeds up to 500 miles per hour."[42]

Others saw the altitude tunnel as a vital research tool. Just days before the Pearl Harbor attack, William Knudsen, Director General of the War Department's Production Management Office, wrote to the NACA Chairman, Jerome Hunsaker, "The development of airplane engines of greater performance at high altitudes is absolutely essential." Referring to the AWT, Knudsen continued, "The high-altitude wind tunnel is especially needed to solve problems in connection with the cooling and power output of engines in combat planes required to fight at altitudes of 40,000 to 50,000 feet."[43]

Carlton Kemper said at the time, "AERL [the new engine lab—the Aircraft Engine Research Laboratory] is unique in having the only altitude wind tunnel in the world. We can expect that this one research tool will give answers to the military services that will more than offset the cost of the laboratory."[44] In the end, Mead would get much of what he sought, in addition to the altitude capabilities proposed by others.

Image 12: Zella Morewitz poses with a model of the NACA Aircraft Engine Research Laboratory, currently the NASA Glenn Research Center. The AWT can be seen near the center of the model. The AWT would be the research and geographical center of the lab for years. Morewitz transferred from Langley with the construction team and remained Dr. Ray Sharp's assistant for a number of years after the engine lab became operational. (NASA C-1942-01009)

Hello, Cleveland

The city of Cleveland put together a detailed proposal to persuade the NACA to build its engine lab in Cleveland. Industrial Commissioner Clifford Gildersleeve and Frederick Crawford, head of both the Chamber of Commerce and Thompson Products, Inc., were the city's principal advocates. Thompson Products manufactured aircraft parts locally, and Crawford himself organized and ran the National Air Races held annually since 1930 at the proposed site for the lab.[45] A crowd of 100,000 had attended the races over the 1939 Labor Day weekend.[46]

The NACA considered 72 bids in 62 other cities before announcing on 15 November 1940 that it had selected the 200-acre area north of the Cleveland Municipal Airport. The property had been used for parking and grandstands for the air races.[47] Proximity to an airport and the ability to generate sufficient

Image 13: Clevelanders swarm to watch the annual National Air Races at the Cleveland Municipal Airport on 5 September 1938. The city offered many advantages for the NACA: Clevelanders had long had an avid appreciation of aviation; Thompson Products was one of many aircraft manufacturing companies in the area; the Case Institute of Technology offered a supply of engineering graduates; and the adjacent airport would allow access for test aircraft. In a little more than two years, construction of the NACA's engine research lab would begin at this site. (NASA C-1991-01875)

electrical power were essential elements. The airport was a given, but the electricity posed a problem. The final hurdle was removed when Crawford negotiated an agreement in which the NACA would operate the AWT and other facilities requiring heavier power loads during the night. In exchange, the Cleveland Electric Illuminating Company promised sufficient power at reasonable rates.[48] This practice continues to this day.

During a luncheon at the Hotel Cleveland immediately before the 23 January 1941 groundbreaking for the Cleveland lab, Crawford reminded Cleveland manufacturers of their promise to supply the new lab with "anything it needs, more promptly, more cheaply, more accurately and more satisfactorily than it can be produced anywhere else."[49] John Victory, who had become friends with Gildersleeve by this time, said, "There seems to be something in the spirit of the people in Cleveland that makes effective cooperation seem easy."[50]

Endnotes for Chapter 1

2. Douglas H. Robinson, *LZ 129 "Hindenburg"* (Fallbrook, CA: Aero Publishers, Inc., 1964), pp. 19–20.

3. U.S. War Department, "Final Report of War Department Special Committee on Army Air Corps" (Washington, DC: Government Printing Office, 18 July 1934), p. 10.

4. George W. Lewis, "Report on Trip to Germany and Russia: September-October 1936," *Historical Collection File 11059* (Washington, DC: NASA Headquarters, 1936).

5. Richard P. Hallion, "America and the Air and Space Revolution: Past Perspectives and Present Challenges" National Aeronautical Systems and Technology Conference (13-15 May 2003), pp. 12–16 and 21.

6. Lewis, "Report on Trip to Germany and Russia," p. 16.

7. Lewis, "Report on Trip to Germany and Russia," pp. 6 and 11–18.

8. "The New Pressure-Type Wind Tunnel at Göttingen," File 11059, NASA Headquarters Archives.

9. Lewis, "Report on Trip to Germany and Russia," pp. 6 and 11.

10. Donald D. Baals and William R. Corliss, *Wind Tunnels of NASA,* Chap. 1 (Washington, DC: NASA Headquarters, 1981).

11. James R. Hansen, *Engineer in Charge: A History of the Langley Aeronautical Laboratory, 1917–1958* (Washington, DC: NASA SP–4305, 1987), Chap. 3.

12. H. Denthloff and H.L. Snaples, *Who Was Albert F. Zahm?* (Reston, VA: AIAA A00–16862, 2000), p. 6.

13. Jakob Ackeret, *Recent Experiments at Göttingen Aerodynamic Institute* (Washington, DC: NACA TM–323, 1925), p. 8.
14. Jerome Hunsaker, "Forty Years of Aeronautical Research" *Smithsonian Report for Aeronautics* (Washington, DC: Smithsonian Institution, 1956), p. 243.
15. Ludwig Prandtl, *Göttingen Wind Tunnel for Testing Aircraft Models* (Washington, DC: NACA TN–66, 1920), pp. 1–6.
16. Prandtl, "Göttingen Wind Tunnel," pp. 2–6.
17. Hunsaker, "Forty Years of Aeronautical Research," p. 256.
18. Baals, *Wind Tunnels of NASA,* Chap. 2.
19. "A Visit to the Langley Field Conference," *Scientific American* (August 1936).
20. Baals, *Wind Tunnels of NASA,* Chaps. 2 and 3.
21. Interview with Benjamin Pinkel, conducted by Virginia Dawson, 4 August 1985, NASA Glenn History Collection, Oral History Collection, Cleveland, OH.
22. NACA, "Thirty-Second Annual Report of the National Advisory Committee for Aeronautics" (Washington, DC: Government Printing Office, 1946), p. 15.
23. Alex Roland, *Model Research,* Volumes 1 and 2 (Washington, DC: NASA SP–1403, 1985), chap. 7.
24. Interview with Abe Silverstein, conducted by Walter T. Bonney, 21 October 1972 and 20 September 1973, NASA Glenn History Collection, Oral History Collection, Cleveland, OH.
25. Lewis, "Report on Trip to Germany and Russia," p. 16.
26. Theodore Von Karman, H.S. Tsien, Hugh Dryden, et al., "Technical Intelligence Supplement: A Report for the AAF Scientific Advisory Group" *Toward New Horizons Supplement* (Dayton, OH: Material Air Command, May 1946), p. 99.
27. National Park Service, "Wright-Patterson Air Force Base Area B Buildings 25 and 24: Photographs, Written Historical and Descriptive Data, Reduced Copies of Drawings," *Historic American Engineering Record No. OH–79–AP,* pp. 3–4.
28. Charles Lindbergh to Joseph Ames, 28 November 1938, File 1308, NASA Headquarters Historical Collection, Washington, DC.
29. Joseph Ames to Charles Lindbergh, 7 December 1938, File 1308, NASA Headquarters Historical Collection, Washington, DC.
30. Roland, *Model Research.*
31. John F. Victory, "Remarks at Cleveland Chamber of Commerce Luncheon Commemorating the Tenth Anniversary of Foundation of the Lewis Flight Propulsion Laboratory," 23 January 1951, NASA Glenn History Collection, Cleveland, OH.
32. NACA, "Twenty-Fifth Annual Report of the National Advisory Committee for Aeronautics" (Washington, DC: Government Printing Office, 1940), p. 4.
33. "Report of the Special Committee on New Engine-Research Facilities" (24 January 1940), File 11059, NASA Headquarters Historical Collection, Washington, DC.
34. "NACA Executive Committee Minutes" (7 February 1940), File 11059, NASA Headquarters Historical Collection, Washington, DC.
35. Charles Lindburgh to Joseph Ames, 4 November 1938, File 1308, NASA Headquarters Historical Collection, Washington, D.C.

36. Dik Daso, "Origins of Air Power: Hap Arnold's Command Years and Aviation Technology, 1936–1945," *Aerospace Power Journal* (Fall 1997): 99 and 110.
37. John Victory to Jerome Hunsaker, "Origin and Status of the Aircraft Engine Research Laboratory," 7 October 1941, File 11059, NASA Headquarters Historical Collection, Washington, DC.
38. Jakob Ackeret, *High Speed Wind Tunnels* (Washington, DC: NACA Report 808, 1925), p. 2.
39. Lawrence Hawkins, "New Aircraft Laboratory Here Will Tackle Knotty Problems Now Baffling Research Engineers," *Cleveland Plain Dealer* (23 February 1941).
40. Ed Clarke, "Engine Laboratory in Hands of Man Who Does Job Well," *Cleveland Press* (27 November 1940).
41. "NACA Wind Tunnel Is In Operation," *Electrical Production Magazine* (August 1944).
42. George Mead to George Lewis, 13 November 1939, File 1292, NASA Headquarters Historical Collection, Washington, DC.
43. William S. Knudsen to Jerome Hunsaker, 2 December 1941, File 11059, NASA Headquarters Historical Collection, Washington, DC.
44. "Stresses Need for Speed in Research," *Wing Tips* (25 March 1944).
45. John D. Holmfeld, "The Site Selection for the NACA Engine Research Laboratory: A Meeting of Science and Politics" (Bachelor's thesis, Case Institute of Technology, 1967), 40.
46. Tom Huntington, "Racetracks in the Sky," *Invention and Technology* (Spring 2007): 51.
47. "City Gets $8,400,000 Plane Lab," *Cleveland Plain Dealer* (25 November 1940).
48. Interview with Frederick Crawford, conducted by Tom Farmer, *This Way Up: Voices Climbing the Wind WVIZ Documentary*, 1991, NASA Glenn History Office, History Collection.
49. "Obligations on Cleveland," *Cleveland Press* (24 January 1941).
50. "Credits Winning of Plane Lab to City's Cooperative Spirit," *Cleveland News* (2 December 1940).

Image 14: The Altitude Wind Tunnel as its shell is constructed by the Pittsburgh-Des Moines Steel Company in April 1943. (NASA C–2008–00817)

Chapter 2

Building a New Type of Tunnel | Design and Construction (1940–1943)

Design work for the new Aircraft Engine Research Laboratory (AERL) was well under way at NACA Langley by the time of the 23 January 1941 ground-breaking ceremony in Cleveland. Under the guidance of Ernest Whitney, the men hunched intently over drawings and calculations in a room above the new Structural Research Laboratory. They labored, sometimes two or three to a single desk, with the difficult task of designing complex engine test facilities. Initially the engineers did not know the location of the new engine research lab, let alone other important criteria, such as where the utilities would tie in.[51]

The new lab would have six principal buildings: the Engine Research Building, a hangar, the Fuels and Lubricants Building, the Administration Building, the Propeller Test Stand, and the Altitude Wind Tunnel (AWT), formerly referred to as the Engine Research Tunnel. U.S. involvement in World War II was becoming increasingly unavoidable, and there was tremendous pressure to get the AERL operating. Both the army and navy had a backlog of aircraft engine problems that needed to be resolved. Ironically, it was the war that slowed the construction progress down. Extraordinary measures would have to be implemented to get the lab up and running in time.

The massive wind tunnel was the key component in the overall design of the new lab and would be its greatest engineering challenge. The *Cleveland Plain Dealer* reported that the AWT would require more engineering man-hours than the Boulder Dam.[52] The 263-foot-long by 120-foot-wide

AWT would include six auxiliary buildings. The AWT's altitude environment required enormous refrigeration and exhaust systems and a structure robust enough to withstand the associated strain. The tunnel would be capable of testing engines twice the horsepower of any engine in operation at the time. In order to run the engines within the closed-loop tunnel, an exhaust scoop would have to be developed to pump the polluted air out of the tunnel. This lost airflow would have to be simultaneously replenished, and the engines would have to be operated, measured, and fueled.

Designing a Wind Tunnel for Engines

The main AERL design group at Langley consisted of approximately 30 engineers and draftsmen, but there were smaller groups working separately on specific facilities. Among these was a group led by Larry Marcus and Al Young that planned the AWT's exhaust and makeup air systems, refrigeration, control room, test chamber, and support buildings.[53] Some elements of the plans had been used for the most recent Langley tunnels, but others—such as the air scoop and cooling system—were new.

Image 15: The AERL design team works in an office above the Structural Research Laboratory at NACA Langley in April 1941. Less than eight months later, they would be transferred to Cleveland with other Langley personnel to complete the design work on site. This group included future Cleveland leaders, including Addison Rothrock, George Darchuck, Harold Friedman, and Nick Nahigyan. (1941) (NASA C-2007-02563)

Another group, which had been transferred to the new NACA Ames Aeronautical Laboratory, undertook the task of designing the AWT's shell and electrical drive system. Carlton Bioletti led the team, which included Walter Vincenti, John Macomber, and Manfred Massa. Bioletti had recently put Langley's 19-Foot Pressure Tunnel into operation and was in the process of designing the drive system for the new 40- by 80-foot tunnel at Ames. The AWT engineers were harried since they were also working on the new wind tunnels at NACA Ames.[54]

The overall layout of the AWT was similar to that of other NACA wind tunnels, such as Langley's 8-Foot High Speed Tunnel and 19-Foot Pressure Tunnel, but the temperature and pressure fluctuations due to the altitude simulation made the design of the shell more difficult than for previous tunnels. The simultaneous decrease in both pressure and temperature inside the AWT would produce uneven stress loads, particularly on the support rings. Since this was an entirely new type of engineering problem, Vincenti took his best guess at how to calculate for it and then consulted with his former professor at Stanford, Stephen Timoshenko. Timoshenko, a leading expert on

Image 16: The 1-inch-thick inner shell of the AWT is visible in this photograph of the facility being erected. The shell was made from a special steel alloy similar to current ASTM A710 Grade A3 steel plate. It was significantly stronger than normal carbon steel. (1943) (NASA C-2007-02314)

structural dynamics, developed some calculations that addressed the problem, and Vincenti forwarded the calculations to the main design team. Although they found them incomprehensible at first, it was not long before they grasped Timoshenko's insightful method of analysis.[55]

The AWT's shell would be forged from a steel alloy capable of withstanding low temperatures. The steel was 1 inch thick to ensure that the shell did not collapse as the internal air pressure was dropped to simulate high altitudes. It was a massive amount of steel considering the wartime shortages. The shell was to be covered with several inches of fiberglass insulation to retain the refrigerated air and a thinner outer steel layer to protect against the weather. A unique system of rollers was used between the shell and its support piers. These rollers allowed for movement as the shell expanded or contracted during the altitude simulations. The section near the refrigeration system would move as much as 5 inches during operation.[56]

The Birth of Cool

One of the AWT's most daunting engineering requirements was cooling the millions of cubic feet of airflow per minute. The refrigeration system would have to cool the airstream and remove the substantial amount of heat being produced by the tunnel's drive fan and the engine being tested.[57] The Langley team devised a refrigeration system that employed a new type of cooling coil with streamlined tubes. By November 1941 it was clear that Langley's development of the refrigeration system was lagging behind that of the AWT's other components. The NACA decided to consult with Willis Carrier, whose Carrier Corporation had pioneered the field of air conditioning and refrigeration.

AERL leaders, including Al Young, Lou Monroe, Dr. Edward "Ray" Sharp, and Rudolph Gagg, met with Willis Carrier and his representatives on 6 November 1941 to discuss the project. The NACA contingent left feeling "impressed by the confidence with which Carrier approached this problem; [the Carrier representatives] seem entirely capable of carrying out a project such as ours."[58] Willis Carrier agreed to bid on the job. He later claimed that he informed George Lewis that "the [Langley] boys conducting the tests did not know what it was all about, and that too much money, and of more importance, too much time had been wasted already."[59]

Willis Carrier created several teams at his Buffalo, New York, plant to work on different aspects of the project. Maurice Wilson managed the engineers,

Samuel Anderson designed the cooling coils, and Adolph Zulinke created the special refrigerant controls, but it was Carrier himself who was the driving force on the project.[60] Once the contract was signed, Carrier had two to three months to complete the design work. Anderson recalled that the team often put in 16 to 18 hours per day on the project.[61]

Anderson and Everett Palmatier initially attempted to use a standard heat exchanger setup. When all of the cooling coil tube designs failed to meet the pressure drop called for by the NACA, a new design was attempted. To overcome the lack of adequate surface area for the heat exchangers, the Carrier engineers decided to install the cooling coils in a folded accordionlike arrangement that provided 8,000 square feet of surface area. They would build a full-size version of one-half of a heat exchanger to test the design.[62] A tunnel section was erected over the exchanger so that air could

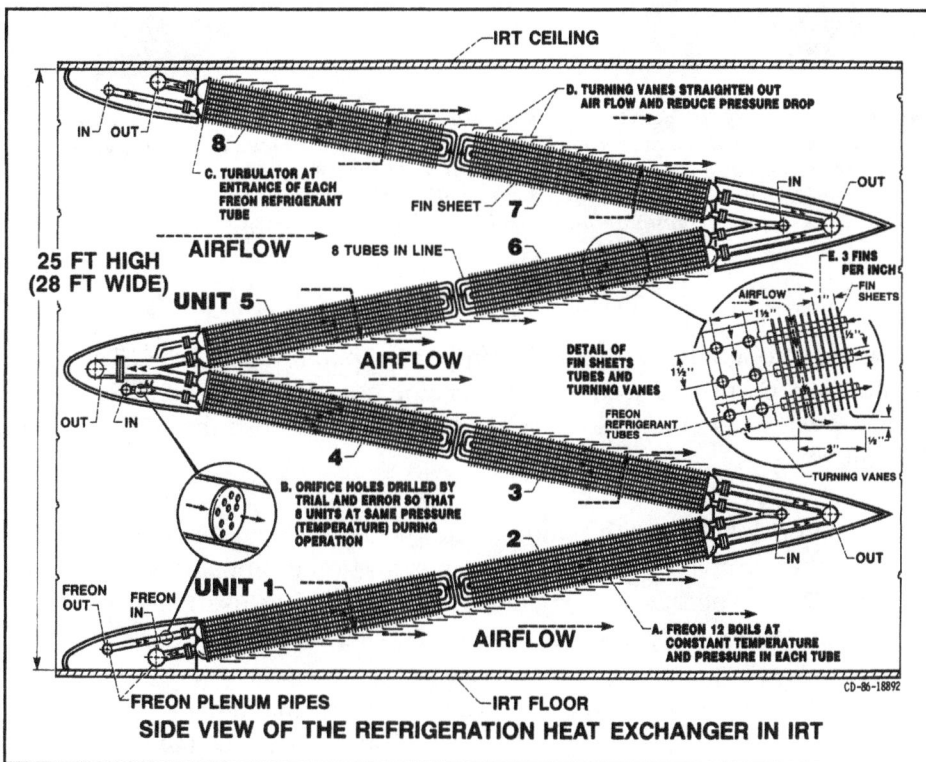

Image 17: Side view of the heat exchanger setup in the Icing Research Tunnel. This arrangement was almost identical to that for the larger heat exchangers in the AWT. The accordionlike design was created by the Carrier Corporation to increase the surface area of the cooling coils eight times. (NASA C–1956–41911)

Image 18: Interior of the Refrigeration Building showing the 14 Carrier Corporation 1,500-horsepower centrifugal compressors that were the backbone of the AWT's complex cooling system. The system was used to cool the airflow in both the AWT and the Icing Research Tunnel. The building also contained York compressors, which were used to refrigerate the AWT's makeup air and provide institutional chilled water for the laboratory. (NASA C-1944-7456).

be passed through the device at the proper temperature and scale. The setup would allow Carrier's engineers to determine not only the optimal size of the exchanger, but also the amount of force required to push the wind through the coils.[63]

Rather than risk failure with standard equipment, Carrier engineers designed many of the pumps, valves, flexible joints, and other apparatus specifically for the AWT project. Carrier engineers also decided to use Freon-12 as the refrigerant. Although Freon-12 would be common in the coming years, the use of large quantities of it had not been attempted previously. Carrier developed a method of circulating the Freon at quantities and pressures large enough to ensure complete distribution throughout the coils.[64] The engineers were having difficulty with the decrease in speed after the airflow exited the coils. This was resolved by installing turning vanes across the back end of the coils to make the cold airflow slightly faster than the tunnel's normal airflow.[65]

The success of the AWT's cooling system was one of Willis Carrier's greatest accomplishments. He had kept a close eye on the project and had made several key suggestions himself, including the turning vanes.[66] The system was powerful enough to also cool the Icing Research Tunnel—a smaller atmospheric tunnel located just behind the AWT—and provide institutional chilled water for the lab. After over 60 years of operation, the system remains in use today.

Struggling To Build the Laboratory

The creation of the NACA Ames lab had progressed on schedule and within its budget. The first Ames technical report was issued in April 1941: just 15 months after ground had been broken.[67] The nation had changed a great deal during the 13-month interlude between the groundbreaking events in Sunnyvale and Cleveland. Resources and funding became more difficult to

Image 19: Center and right: The AERL's first two employees—Construction Engineer-Inspector Charles Herrmann and administrative assistant Helen Ford—arrived in Cleveland from Langley in early February 1941, setting up offices in a "radio house." On 30 July they relocated to the farm house, seen in this photograph, which had been acquired with the property. This farm house served as the original administrative building as the AERL was being constructed. (1941) (NASA C–2006–01209)

obtain as war approached.[68] In addition, Langley engineers had less experience with the AERL's engine research facilities than with the aerodynamic facilities built at Ames. This meant that consultants from Wright Field, engine manufacturing companies, and other companies, such as Carrier, had to be brought in to assist.

The first Langley personnel began arriving in Cleveland in February 1941. There were still no buildings completed when Dr. Ray Sharp transferred from Langley six months later to personally oversee the construction. Sharp was soon versed on almost every construction project at the lab. The nation would enter the war before the end of the year, and wartime production needs were in direct competition with the AERL for limited resources. The war, therefore, was the cause of many of the construction delays. Materials were difficult to obtain, contractors were overburdened, and funding was tight.[69] A glimmer of progress appeared with the signing of the construction contract for the AWT just as the year ended.

Late in the day on 31 December, Sharp traveled to Washington, DC, to discuss the Sam W. Emerson Company construction contract with NACA Secretary John Victory. After the details were finalized, the two walked the $95,000 contract over to the White House, where it was approved by President Franklin Roosevelt at 6:30 p.m.[70] Emerson was now the primary construction contractor, in charge of building the AWT's foundations, the Shop and Office Building, and the support buildings.[71]

Image 20: Dr. Ray Sharp had briskly moved through the ranks at Langley before being named as the AERL's Officer and Construction Manager. In March 1940, Sharp was detailed from Langley to supervise the construction at Ames until Smith DeFrance assumed control in August. Afterward, Sharp and construction engineer Ernest Whitney traveled to Cleveland to perform a property survey as part of the site-selection process for the new engine lab.[72] Shortly thereafter, Sharp was recalled to Langley and named Chief of the Construction Division. In August 1941, he was detailed to Cleveland to oversee the construction. He would spend the next 20 years leading the Cleveland lab.[73,74] (NASA C-1943-01534)

Emerging from a Sea of Mud

A large contingent of Langley personnel led by Construction Supervisor Beverly Gulick and supervising engineer Ernest Whitney arrived in December 1942. A month later the hangar became the first building completed. When the Engine Propeller Research Building was completed in May, a ceremony was held to initiate research at the lab. The media and NACA officials were on hand when George Lewis activated a 14-cylinder R-2600 Wright Cyclone engine.[75]

Most of the lab was still a sea of mud, however. Excavation for the AWT's foundations had begun only recently. Just three days after the ceremony, General Henry "Hap" Arnold requested that the NACA's priority rating be elevated to Class 1 to expedite the allocation of resources. George Lewis

Image 21: This photograph from late spring 1942 shows the construction of the AWT's Shop and Office Building. The tunnel's steel and concrete piers can be seen behind the crane. The Sam W. Emerson Company would complete this task by late December. The hangar, the first AERL building completed, is in the background. Retiree Frank Holt recalled that during this period the surveyors were "out everyday in their high-top boots with their tripods and scopes" in the mud laying out the new lab.[76] (1942) (NASA C-2007-02306).

began arriving from Washington every Monday to oversee the lab's progress.[77] He stated at the time, "Originally planned for completion over a period of two and one-half years, this new NACA laboratory is being rushed to completion at least one year ahead of schedule because of its importance in [the] present war effort."[78,79]

In September 1942, U.S. Army Air Force Major General Oliver Echols wrote, "The work of the NACA is a vital and inseparable part of the aircraft production program. Much depends upon its ability to be fully ready by

Image 22: Left to right: Ray Sharp and George Lewis speak to AERL employees in May 1942. Construction was progressing but would be expedited in the coming months in an effort to have the lab operational by the end of the year. Lewis began visiting the site weekly to personally assess its progress. Work on the AWT's support buildings and tunnel foundation had started, but the majority of its construction would take place in 1943. (NASA C-1942-08287)

December 31, 1942, to meet growing demands of the military services. I must therefore urge that the NACA be given the highest preference in obtaining needed materials and equipment."[80]

Drastic measures were undertaken to accelerate the lab's construction schedule. The military provided special supplies, contractors were given new agreements and pressured to meet deadlines, and Congress approved additional funds.[81] Entire sections of NACA Langley engineers and researchers, including the Powerplants Division, were transferred to the AERL.[82]

Image 23: Left to right: NACA Secretary John Victory, U.S. Navy Fleet Admiral Erwin King, and AERL Manager Ray Sharp. In a 14 July 1942 letter to the Army and Navy Munitions Board, King wrote, "The present A-1-A priority rating is not high enough to assure its readiness in time. Whatever higher priority rating or special directive may be necessary to avoid delay in its completion is justified and recommended."[83] (NASA C-1945-11620)

Model Builders Prove Their Worth

"Hurry boys and girls, make your hobby pay you back in money and satisfaction of helping to do your part. Help our planes fly higher, better and faster than those of the Axis by aiding the research engineers," beckoned a May 1944 "Junior Aviators" column in the *Cleveland Press*. [84]

A large contingent of Langley construction personnel arrived at the AERL in December 1942. These veteran engineers would be the foundation for the new lab's research and management staffs. Most of the mechanics and technicians, however, were journeymen from the Cleveland area. There was a need for additional workers, and the AERL appealed to young people in high school.

Frank Holt, a retired technician, recalled in 2005 that the NACA and AERL Construction Manager Ray Sharp were particularly interested in model builders. Beginning at Langley and carrying over to Cleveland, Sharp sought model builders to craft the large wooden propellers for the wind tunnels. After overcoming initial opposition from veteran NACA craftsmen, the young model builders demonstrated their skill and were soon hired for a variety of permanent positions. [85]

The nation had just entered the war when Holt, then just 16 years old, read about the new engine laboratory in one of the "Junior Aviators" columns. He was an avid model builder who had won several prizes, so he decided to send in an application. "I was amazed when they called me," he remembered.

Image 24: Hired at the age of 16, Frank Holt was likely the youngest employee to ever work at the lab. He was one of the many technicians hired in the lab's initial years because of his model-building skills. After the war he crewed a team that won two trophies in the National Air Races held annually in Cleveland. (NASA C–1943–01331)

He recalled going to the AERL employment office in a small radio shack at the edge of the construction site on a blistery January morning. "I came into the room, and there was a bunch of guys I recognized as top model builders. I felt very inferior. I didn't have a chance." During the interview, Holt's mother, who had been shivering outside in the car, came inside to warm up, mortifying the young applicant in front of his peers. Despite his self-consciousness, the interview went well and Holt began work on 9 February 1942. He would spend the next 38 years working in the lab's wind tunnels.[86]

Model builders had two roles: building scaled versions of facilities and creating aircraft test articles that could be studied in the wind tunnels. Since the AERL would have no tunnels in operation until 1944, the early model builders worked on teams, creating the elaborate facility models piece by piece, or were reassigned to new positions as mechanics in the hangar.

The AERL's recruiting campaign continued throughout the war. NACA literature told prospective employees that following World War I and through the Depression the NACA had not laid off a single employee because of a lack of funds. "Unlike industry, the laboratory has no products to sell. Its purpose is not to make money. Research on aircraft engines will go on whether industry turns out one or one million planes."[87] Compensation was not the greatest, however. In 1942 AERL beginning model builders earned less than $1,300 annually, with increases as they worked their way up the ranks.[88]

Image 25: This wooden scale model of the Bell YP-59A Airacomet was created by the AERL's model shop in 1944. It was used as part of the first test in the new AWT. For the test, an actual full-scale fuselage with its two jet engines was installed in the tunnel test section. (NASA C-1945–11554)

Image 26: NACA construction engineers Lou Hermann and Jack Aust assemble the AWT drive fan inside the hangar at the AERL. This 12-bladed, 31-foot-diameter spruce wood fan was designed at NACA Langley. John Breisch, a Langley technician with several years of wind tunnel installation experience, was brought up in July to supervise the fan assembly. He would return several weeks later to oversee the actual installation in the tunnel.[89] (NASA C-1943-01849)

Image 27: On 7 July 1943 the Memphis Belle and its crew visited the AERL as part of a publicity tour. The B-17 Flying Fortress and its crew were returning after 25 successful bombing missions over Germany. Inside the hangar in the background, the large fan for the AWT was being assembled at the time. (NASA C-1943-1867)

Photo Essay 1:

The Altitude Wind Tunnel Stands Up

Image 28: Corner rings for the AWT are raised into place on 9 January 1943 during the early phases of the shell's construction. The Pittsburgh-Des Moines Steel Corporation constructed most of the enormous 263-foot-long, 120-foot-wide rectangular tunnel. The ring in the foreground had a 51-foot interior diameter. (9 Jan. 1943) (NASA C-1944-06707)

Image 29: Steel framework for the Shop and Office Building is at the left, with one of the tunnel's corner rings standing vertically. The completed aircraft hangar is in the background, and the tunnel's foundations are in the foreground (viewed from the southwest). (1942) (NASA C-2007-02294)

Image 30: Construction of the AWT with its test chamber visible in the center. Construction of the tunnel shell began in early 1943 and was completed in January 1944. In this photograph, the outer layer of the shell had yet to be installed (viewed from the south). (NASA C-1943-01521)

Image 31: Interior of the Refrigeration Building with the flash cooler in the center of the photograph. The flash cooler would be connected to the distributing headers and the Carrier compressors along both the left and right. (NASA C–2007–02318)

Image 32: During July 1943 the AWT's drive motor was installed in the northwest corner of the Exhauster Building. The motor, whose support frame is seen in this photograph, connected to the drive shaft that extended from the building, through the tunnel shell and into the AWT's fan assembly. The 18,000-horsepower General Electric Company induction motor was installed, and the corner of the building was built around it afterward. (1943) (NASA C–1943–01962)

Image 33: Construction of the viewing platform in the AWT's Shop and Office Building. The platform is in the rear of the high bay. The shop can be seen below to the right. The lower half of the tunnel's test section, out of view to the left in this photograph, was sunken below the platform so that the test articles were near the floor level. (NASA C-2007-02298)

Image 34: Interior of the shop area inside the western wing of the AWT's Shop and Office Building as it was being constructed. (NASA C-1943-02186)

Image 35: Interior of the AWT's southern leg during construction. The propeller bearing supports are in place, and the bearing box propeller hub and drive shaft are being installed at the far end. (1943) (NASA C-1944-04710)

The Final Push

The frame of the AWT's Shop and Office Building was complete, and excavations had begun for the Exhauster and Refrigeration buildings. The AERL staff worked 48-hour weeks throughout 1942, but construction of the actual tunnel was only beginning when the lab closed for its only holiday that year, Christmas Day.[90] As the AERL opened for business in 1943, construction of the AWT went into high gear. The Pittsburgh-Des Moines Steel Company was responsible for building the tunnel structure. They would erect the remainder of the support rings and tunnel shell and would install the exhaust scoop, turning vanes, and intake air vents during 1943.[91]

The support buildings were largely completed by September 1943, but elements of the refrigeration and exhaust systems required additional time to install.[92] Work proceeded on the balance chamber and test section, including the test section's hinged door. Work on the tunnel, test section, balancing scales, and fan continued into the new year.

Image 36: The test section lid, seen raised in this photograph, was 40 feet long, 20 feet wide, and 10 feet high. A motor-driven system using large counterweights, pulleys, and cables could open and close the lid in approximately 10 minutes. The lid contained a number of viewing windows and a portal for a periscope camera. Handwheels were used to seal the lid once it was lowered in place. (NASA C-1950–26294)

Harold Friedman, a young engineer who had recently finished a project, approached the construction supervisor, Ernest Whitney. He recalled saying, "'Mr. Whitney, I don't have a goddamned thing to do. What should I do?' Ernest told me that I should go over to the Altitude Wind Tunnel, which was under construction, and talk to Al Young and Lou Monroe because they needed somebody to design a door." "I said, 'Sure.' What the hell's so hard about designing a door?" Friedman's "door" was a massive mechanical clamshell-shaped lid that sealed the tunnel's test section. It used large counterweights, pulleys, and cables to open, close, and lock in place.[93]

The Most Important Component

In January 1943, the AWT and the Icing Research Tunnel joined the Engine Analysis, Engineering Drafting, and Unconventional Aircraft Engine Research groups in the Engine Installation Division.[94] Abe Silverstein, the 35-year-old Chief of Langley's Full-Scale Tunnel, was selected by George Lewis to manage the AWT. Silverstein was in the aerodynamics group but had always had a strong interest in aircraft engines. Silverstein later recalled Lewis informing him of the reassignment, "I can see that you want to be in the

Image 37: Left to right: Abe Silverstein and Smith DeFrance. Silverstein transferred to Cleveland after 14 years at Langley. He had played a key role assisting deFrance with the design of the Full-Scale Tunnel. He served as chief of the facility from 1940 to 1943. Although he mainly cleaned up drag problems on fighter aircraft, he did sneak in two engine tests during this period. (NASA C-2009-02181)

engine business, and I want to give you the chance."[95] Most of the Langley personnel previously transferred to Cleveland were from the construction and powerplants groups. At Langley, Silverstein did not have the best relationship with the powerplant people, so there was some tension when he arrived just in time to manage the new world-class facility. Lewis escorted Silverstein to Cleveland in the fall of 1943 to make a formal introduction and smooth things over. After that initial meeting, Silverstein claimed that he had never had any serious problems with the original Langley crew.[96]

Silverstein immediately began assembling the 90-person AWT section. Not surprisingly, he chose individuals such as Al Young, Lou Monroe, and Harold Friedman, who had been involved in the design and construction of the tunnel. The original AWT organization consisted of 40 engineers, 25 mechanics, 15 analysts, and 10 computers.[97] "Computers" were female employees hired to record raw test data and perform mathematical calculations in order to make the data useful.

Image 38: Al Young was involved with the AWT from its design at Langley to his retirement in 1970. For most of that time, he managed the tunnel's operation. Silverstein remembered him as "a superb mechanical engineer, absolutely able." The two shared an office in the AWT for almost four years. Silverstein recalled, "[Young] never spoke unless he was very sure what he was going to say...he was very good and I enjoyed working with him."[98] (NASA C–2008–00629)

Image 39: The AWT complex as it originally appeared. The Shop and Office Building is in the center, the Refrigeration Building is to the right, and the Exhauster Building to the left. When completed in early 1944, it became the first wind tunnel in the nation designed specifically to study engine behavior and capable of creating a high-altitude environment. (NASA C–1945–13045).

Silverstein shared room 103 in the Shop and Office Building with his deputy, Young; Friedman; and Monroe, an ex-Carrier employee who oversaw the installation of the refrigeration system. Other members of the design team, including John Macomber and Manfred Massa, were located upstairs. With the exception of Young, these men would go on to design the 8- by 6-foot and 10- by 10-foot supersonic wind tunnels in Cleveland. Other original AWT staff members included future leaders such as G. Merritt Preston, Robert Godman, Myron Pollyea, DeMarquis Wyatt, and Austin Reader.[99]

The core members of the AWT's long-term research staff were assembled by early 1944. The original team included future manager Al Young, principal AWT engineers such as William Fleming, Robert Dietz,[†] and Martin Saari, as well as James Quinn, the head of the AWT mechanics, and Clifford Talcott, who managed the AWT's electrical supply.[100]

[†]Years later, Dietz would be a key advocate of NASA's most recent wind tunnel, the National Transonic Facility at Langley.

After almost two years of construction, the AWT was finally completed in January 1944. Willis Carrier and his team were on hand for the AWT's initial trial runs to ensure that the refrigeration system worked properly. The first official test took place on 4 February 1944.[101] For the next year and a half, the AWT would contribute to not only the current war effort, but more importantly to the future development of aircraft engines. Although progress seemed to lag because of the urgency of the war, the construction of the AWT in less than two years was a remarkable feat. In the end, the new NACA lab was completed ahead of schedule but at nearly twice the estimated cost.[102]

Endnotes for Chapter 2

51. Interview with Harold Friedman, conducted by Bob Arrighi, 2 November 2005, NASA Glenn History Collection, Oral History Collection, Cleveland, OH.
52. "City Gets $8,400,000 Plane Lab," *Cleveland Plain Dealer* (25 November 1940).
53. Ernest G. Whitney, "Lecture 22—Altitude Wind Tunnel at AERL," 23 June 1943, NASA Glenn History Collection, Cleveland, OH.
54. Friedman interview, conducted by Bob Arrighi, 2 November 2005.
55. Interview (notes) with Walter Vincenti, conducted by Glenn Bugos, 11 May 2007, Moffett Field, CA.
56. Everett P. Palmatier, "Notes on Refrigeration System for Cleveland Wind Tunnels" (c1993), NASA Glenn History Collection, Cleveland, OH.
57. Carlyle Ashley, "Cooling the NACA (NASA) Wind Tunnels at Cleveland During WWII," NASA Glenn History Collection, Cleveland, OH.
58. Russell G. Robinson to George Lewis, "Conference With Representatives of Carrier Corporation, Langley Field, November 6, 1941" 7 November 1941, NASA Glenn History Collection, Cleveland, OH.
59. Margaret Ingels, *Willis Haviland Carrier: Father of Air Conditioning* (Garden City, NY: Country Life Press, 1952), pp. 97–101.
60. Ashley, "Cooling the NACA (NASA) Wind Tunnels."
61. Samuel Anderson, "Personal Comments of Samuel W. Anderson Engineering Program," NASA Glenn History Collection, Cleveland, OH.
62. Ingels, *Willis Haviland Carrier.*
63. Robinson, "Conference with Representatives of Carrier Corporation."
64. Ingels, *Willis Haviland Carrier.*
65. Ashley, "Cooling the NACA (NASA) Wind Tunnels."
66. Anderson, "Personal Comments of Samuel W. Anderson Engineering Program."
67. Erwin P. Hartman, *Adventures in Research: A History of Ames Research Center 1940–65* (Washington, DC: NASA SP–4302, 1970), chap. 6.
68. Ed Clarke, "US Selects Airport Site," *The Cleveland Press* (25 November 1940).

69. Alex Roland, *Model Research,* Volumes 1 and 2 (Washington, DC: NASA SP–1403, 1985), chap. 7.

70. Edward R. Sharp, "Excerpt From Chronological Record of Events Relative to Cost-Plus-a-Fixed-Fee Contract Naw-1425," NASA Glenn History Collection, Cleveland, OH.

71. "Contract Between the NACA and the Sam W. Emerson Co. for Construction of Several Units of the AERL. NA–1425," NASA Glenn History Collection, Cleveland, OH.

72. "Work on Plane Lab to Start Tomorrow," *The Cleveland Press,* 28 November 1940.

73. Henry Reid to staff, "Reorganization of the Construction Division," 20 September 1940, NASA Glenn History Collection, Cleveland, OH.

74. Henry Reid to staff, "Jet Propulsion Power Plants 1943–48," 20 September 1940, NASA Glenn History Collection, Cleveland, OH.

75. "Initiation of Research" (8 May 1942), press packet, NASA Glenn History Collection, Cleveland, OH.

76. Interview with Frank Holt, conducted by Bob Arrighi, 9 August 2005, NASA Glenn History Collection, Cleveland, OH.

77. Interview with Jesse Hall, conducted by Bonnie Smith, 28 August 2002, NASA Glenn History Collection, Cleveland, OH.

78. Henry H. Arnold and J.H. Towers to Director, Bureau of the Budget, 11 May 1942, NASA Glenn History Collection, Cleveland, OH.

79. Wayne Coy to Henry Arnold, 16 May 1942, NASA Glenn History Collection, Cleveland, OH.

80. Oliver P. Echols to H.E. Talbott, 28 September 1942, NASA Glenn History Collection, Cleveland, OH.

81. James R. Hansen, *Engineer in Charge: A History of the Langley Aeronautical Laboratory, 1917–1958* (Washington, DC: NASA SP–4305, 1987), chap. 8.

82. Helen Ford, "Highlights From Wing Tips," NASA Glenn History Collection, Cleveland, OH.

83. Erwin J. King and G.C. Marshall to Army and Navy Munitions Board, 14 July 1942, NASA Glenn History Collection, Cleveland, OH.

84. Ed Clarke, "Junior Aviators," *Cleveland Press* (22 May 1944).

85. Robert S. Munson, "Pathways Within Model Aviation and Beyond," chap. 1 (2002): *http://www.gryffinaero.com/pathways/* (accessed 29 July 2009).

86. Holt interview, conducted by Bob Arrighi, 9 August 2005.

87. NACA, "NACA Looks to the Future" pamphlet, NASA Glenn History Collection, Cleveland, OH.

88. "Bill Harrison Service Record" NASA Glenn History Collection, Cleveland, OH.

89. George Darchuck, "J. Breisch Aids on AWT Installation, Makes Many Friends Here," *Wing Tips* (30 July 1943).

90. "Highest Priority Work," *Lewis News 25th Anniversary Issue* (23 January 1966) p. 4.

91. "Construction Report No. 145" (November 8, 1943 through November 13, 1943), NASA Glenn History Collection, Cleveland, OH.
92. Aircraft Engine Research Laboratory Construction Report Nos. 66, 100, 106, 134, 145, and 149 (1942–1943), NASA Glenn History Collection, Cleveland, OH.
93. Friedman interview, conducted by Bob Arrighi, 2 November 2005.
94. "See Many Changes in Lab Research Divisions," Wing Tips (5 February 1943).
95. Interview with Abe Silverstein, conducted by Walter T. Bonney, 21 October 1972 and 20 September 1973, NASA Glenn History Collection, Oral History Collection.
96. Interview with Abe Silverstein, conducted by John Mauer, 10 March 1989, NASA Glenn History Collection, Cleveland, OH.
97. NACA Aircraft Engine Research Laboratory, Telephone Directory, 1 September 1944, NASA Glenn History Collection, Cleveland, OH.
98. Silverstein interview, conducted by Walter T. Bonney, 21 October 1972 and 20 September 1973.
99. NACA Aircraft Engine Research Laboratory, Telephone Directory.
100. NACA Aircraft Engine Research Laboratory, Telephone Directory.
101. Ingels, Willis Haviland Carrier.
102. Roland, *Model Research,* chap. 8.

Image 40: The secret test of the Bell YP-59A Airacomet was the first investigation in the new Altitude Wind Tunnel. The Airacomet was the first aircraft in the United States powered by a jet engine, the General Electric I-16. (February 1944). (NASA C-1944-04830)

Chapter 3

Dual Wartime Mission | The World War II Years (1944–1945)

"Will the solution of this problem be of assistance to us in time to be of use to the men on the fighting front?" That was the question Acting Executive Engineer, Addison Rothrock, asked the Aircraft Engine Research Laboratory's (AERL's) new research staff on 20 December 1942 to consider when conducting their investigations.[103] Earlier that day, an engine fire had wrecked the flight test of a Boeing B-29 prototype and grounded the program for seven weeks. The U.S. Army Air Corps had already ordered 1,500 of the long-range bombers and was relying on it to serve as the exclusive weapon for the bombing of Japan.[104] The next flight would end in disaster after another engine fire caused a crash in Seattle that killed 8 crew members and 20 civilians on the ground.[105]

The overheating of the B-29's 18-cylinder radial engines was precisely the type of problem that the Altitude Wind Tunnel (AWT) was intended to resolve. Unfortunately, in December 1942 the tunnel's corner rings were just being put in place. It would be another year before the facility would become operational. Unknown to Rothrock and the audience was that, two months before, a Bell XP-59A fighter had flown using the first jet engine built in the United States. This advanced technology would threaten to make the new AERL facilities obsolete unless they adapted and returned to research.

The dichotomy of the AERL's mission was later emphasized during a visit by General Henry Arnold. Arnold told the assembled staff, "You've got a dual task. You've got a job ahead of you to keep the army and the navy air forces equipped with the finest equipment that you can for this war. You also have the job of looking forward into the future and starting now those developments, those experiments, that are going to keep us in our present situation—ahead of the world in the air. And that is quite a large order, and I leave it right in your laps."[106]

Image 41: General Arnold addresses AERL employees. As Commander of the U.S. Army Air Forces during World War II, he almost single-handedly dragged the nation's aviation leaders into the era of the jet engine, while simultaneously trying to squeeze all of the power from the nation's largest piston engine for his B-29 program. (NASA C-1944-07493)

The NACA's traditional mission of basic research was to be mothballed for the duration of the war. The army and navy relied on the NACA to solve very specific problems with existing military aircraft and their piston engines. The NACA Langley and NACA Ames aeronautical laboratories used their large wind tunnels to reduce the drag of military aircraft and also improved ice prevention and dive bombing. The AERL initially concentrated its efforts on propulsion problems such as engine knock and turbocharger performance. Once the AWT and its sister facility, the Icing Research Tunnel, became operational in 1944, the AERL was able to considerably expand its work. The two tunnels were used for military applications 93 percent of the time during the last year and a half of the war. The Langley and Ames tunnels, which had been active for the entire war, spent 57 percent of their operating time on military tests.[107]

Burning Up in the Sky

One of the most pressing problems facing the air force was the overheating of the Wright R-3350 engines that were used to power Boeing's new long-range bomber, the B-29 Superfortress. Although twice as heavy and powerful as its predecessor—the B-17—the B-29 was designed to fly significantly faster, longer, and higher. It was one of the most formidable and versatile aircraft ever designed and the top priority of the air force. The B-29 was plagued with problems, though, most notably the overheating of the engines as the aircraft reached its higher altitudes.[108]

Missouri Senator Harry Truman led a legislative investigation into the March 1942 crash that had resulted in 29 deaths. The committee concluded that the B-29 crisis stemmed from Wright's poor quality control and the rush of the air force's development schedule.[109] Over a year before the first test flight had taken place, the War Department had outfitted the nation's factories to mass-produce the bombers.[110]

The most significant problem was the engines. The fuel-injection system frequently caught fire, and the combination of poor airflow and engine strain caused the engines to overheat during the B-29's climb to its 30,000-foot cruising altitude. The overheating was exacerbated by the use of magnesium crankcases, which were strong and lightweight but highly flammable.[111] The crew had less than a minute to bail out once the crankcase ignited. The nation's most modern flying machine seemed more like a death trap.

Image 42: A Boeing B-29 participates in the AERL's intensive wartime study of its Wright R-3350 engines. (NASA C–1944–05883)

The bombers were hurried into service in the Pacific and received their testing in combat. The R-3350s would require over 2,000 modifications. In 1944 alone, 54 key revisions had to be implemented on every new B-29.[112] It was not uncommon for the R-3350 to have to be overhauled after less than 100 hours of operation. Yet pilots were still reporting that their engines were overheating and losing power at higher altitudes.[113]

Frank Bechtel recalled in 2004, "They were flying these planes over Japan and back nonstop. They weren't losing any of them to ground fire or Japanese fighter planes, but the planes were going in the drink because the engines were burning up. They [would] get overheated and just quit, and the planes were going in."[114]

Despite the pressure to complete the AWT in order to analyze the R-3350 cooling problems, it was the General Electric (GE) I-16 jet engines on the Bell XP-59A Airacomet that were tested first when the facility was finally ready in early 1944. This decision would prove significant for both the B-29 and the AWT's future focus on the turbojet.

"Nobody Was Really Looking Ahead"

Upon assuming control of the U.S. Army Air Corps in 1938, General Arnold called a meeting to identify vital aeronautical research and development areas. One of the items on the table was the jet-assisted takeoff developed by Frank Malina and Jack Parsons at the California Institute of Technology. Committee members Jerome Hunsaker and Vannevar Bush revealed the NACA's closed-mindedness at the time by deriding the proposal.[115] Three years later Arnold became aware of the turbojet developments in Europe and ordered the NACA to explore the possibilities. Many in the air corps, including NACA member Major General Oliver Echols, urged continued concentration on reciprocating engines.[116]

The success of the turbocharger—which was perfected at GE's West Lynn, Massachusetts, plant—caused U.S. aircraft manufacturers to become complacent. The turbocharger used the engine's hot exhaust gases to spin a turbine that powered a compressor. Because it supplied the engine with additional air, the turbocharger resulted in significantly greater speeds and altitudes for piston aircraft. The Lockheed P-38 and Republic P-47 were considered to be the fastest fighters in the world, but the Boeing B-17 and Consolidated B-24 were the only 30,000-foot-altitude heavy bombers.[117]

The Langley Powerplants Division was just beginning to delve into the turbojet field, even though its staff had been aware of the gas turbine for years. They had felt that, with existing materials and technology, jet engines would not be any more efficient than piston engines. All three of the NACA's studies on the turbojet in the 1930s were farmed out to the National Bureau of Standards.[118] Abe Silverstein later explained that the NACA was still primarily an aerodynamics-based agency and that when it came to propulsion "nobody was really looking ahead."[119]

Other nations were not only looking ahead, but moving ahead. German and British engineers had not mastered the turbocharger, so they had sought other ways to improve engine performance. Engineers in both countries were drawn to the gas turbine. By late 1939, the German government had funded two jet engine design programs and contracted with its two largest aircraft manufacturers to develop fighters that would incorporate the new

Image 43: Left to right: Colonel Edwin Page, Dr. William Durand, Orville Wright, and Addison Rothrock tour the AERL in Cleveland, Ohio, on the lab's May 1943 Dedication Day. Durand, former chairman of the NACA, had been called out of retirement to head the NACA Special Committee on Jet Propulsion. Prior to becoming the AERL's Air Technical Service Command Liaison Officer in May 1943, Colonel Page had spent most of the previous 18 years developing, designing, constructing, and overseeing the research for the engine test facilities at Wright Field. (NASA C-1943-01562)

turbojets.[120] Although they lacked Germany's coordination, the British also were pursuing the jet engine. After a visit to the United Kingdom, Northrop Aircraft's Joseph Phelan related how impressed he was by "the established air of something which had definitely arrived, so to speak. Nowhere did there appear to be anything but a profound belief and faith in the future of the gas turbine in some form or another."[121]

As the Langley engineers were in the initial stages of designing the AWT for piston engine testing, the Europeans were beginning to fly turbojets successfully. In late August 1940 the Italians flew a ducted-fan jet engine, in early 1941 the first rocket-propelled aircraft was flown in Germany, and on 15 May 1941 the British flew their first jet aircraft. While the AWT's foundations were being installed in July 1942, the German Messerschmitt Me-262 Schwalbe, which could fly 540 mph, became the world's first operational fighter jet.[122]

Image 44: Left to right: Henry Reid, NACA Langley Engineer-in-Charge; Carlton Kemper, Chief of the Powerplants Division; and Elton Miller, Chief of the Aerodynamics Division, in an April 1929 division meeting. Neither the Langley powerplants group nor Kemper thought much of the jet engine. Kemper agreed with a 1923 National Bureau of Standards study stating that fuel consumption and weight made jet propulsion impossible. Kemper, who would become Executive Engineer at the AERL, did not reconsider the jet engine until April 1940, and then only in a limited fashion. (1929) (NASA EL–1997–00141)

Although the decision had been made to fight the war with piston aircraft, it was obvious to Arnold and his closest advisors that aviation's future lay with the turbojet. At Arnold's request, the NACA called Dr. William Durand out of retirement in March 1941 to head a Special Committee on Jet Propulsion. The committee—which included members of the NACA committee, scientists such as Hugh Dryden, and industry leaders including GE's Sanford Moss—was tasked with coordinating the nation's secret development of the turbojet.

Heads in the Virginia Sand

Edgar Buckingham of the National Bureau of Standards had conducted studies of a gas turbine in 1923 and concluded in his report for the NACA that the jet engine would never surpass the reciprocating engine's performance. The Langley engine group, headed by Carlton Kemper, accepted Buckingham's findings and showed no interest in the possibilities of gas turbines. Although the conclusions were true at the time, the NACA did not foresee the turbojet's potential for high-speed flight. Silverstein said later, "Unless you visualize airplanes going at 400 to 500 mph, you can't make a point for the gas turbine. And airplanes weren't flying that fast at the time."[123]

Unaware of the German and British advances made during the interim, Langley engineer Albert Sherman replicated the Buckingham study and produced an April 1940 report which concluded that the gas turbine was indeed viable. Sherman's findings convinced Kemper and Langley Engineer-in-Charge Henry Reid that the gas turbine was now worthy of further investigation. Sherman and Eastman Jacobs were tasked with the construction of a ducted fan static test cell, referred to as the "Jeep." Before the Jeep was even completed, Durand's new subcommittee ordered the pair to convert the Jeep setup into a true turbojet capable of being integrated with an airframe. This was a daunting task that lasted the remainder of 1941.[124]

The Jeep's first fiery runs occurred over a year later in February 1942. After a series of improvements, the system was considered to be operational in July. In October, Durand's special committee visited Langley for an exhibition of the new Jeep powerplant. The engine's failure during the big demonstration for all intents killed the program. It was officially ended with the transfer of the Powerplants Division to Cleveland in December.[125] NACA work on axial-flow compressors during the war did influence the manufacturers, but the NACA was left out of the design work for the first wave of turbojets.

Secret Delivery for Cleveland

In April 1941 General Arnold had witnessed the first jet-powered flight in Britain, the Gloucester E-28/39 with its Whittle W-1B engine. Through the Lend-Lease agreement, plans for the Whittle centrifugal engine were secretly brought to the United States so that engineers at GE's West Lynn plant could replicate it.

Work on the new engine, called the I-A, was difficult with only Whittle's drawings as a guide. Colonel Donald Keirn, who oversaw the early U.S. turbojet work, and three British engineers then accompanied the shipment of a Whittle W-1X engine to the United States. The development progressed, but the first attempt to operate the engine in March 1942 was disappointing. After several modifications the 1,250-pound-thrust engine was successfully run on 18 April 1942.[126]

Bell Aircraft Corporation had been contracted the previous October to construct an aircraft that would incorporate the new GE I-A engines. The result was the Bell XP-59A Airacomet.[127] During June of 1942 Whittle himself spent several weeks in Massachusetts assisting with the integration of the engine into the airframe. After a secret cross-country railroad delivery and several days of preparations, the XP-59A made its first flight over Muroc Lake, California, on 2 October 1942.[128]

Image 45: Jet Propulsion Static Laboratory at the AERL. GE's West Lynn team had designed this nondescript two-cell test bed to be built at the AERL to test their new Whittle-based jet engines. An I-A, the first turbojet built in the United States, was secretly brought onto the lab disguised as a supercharger and tested in the Jet Propulsion Static Laboratory in the fall of 1943. The I-A had been superseded by the I-16 by this point. The I-16 would be tested in the AWT just a couple of months afterward. (NASA C–1945–12097)

Although the Airacomet flew, it did not perform well and provided little performance enhancement over the piston version of the aircraft. Its speed of 290 mph was a little over half of what the Messerschmitt Me-262 had achieved several months before.[129] General Arnold, who had been aware of the engine for over a year, expressed his concerns over the slow development of jet aircraft. In a 14 October 1942 letter to NACA Chairman Jerome Hunsaker, Arnold stated that it would be easier to improve the overall aircraft design rather than the engine because the overhaul process required less time. More importantly, he realized that jet propulsion engineers were still rare in the United States and that the I-A engines were not designed specifically for fighter planes.[130]

By January 1943 the West Lynn group had begun work on a 1,650-pound-thrust successor, the I-16. The new engine was more powerful than the I-A, but it added additional weight to the already heavy aircraft. Its speed of 409 mph and altitude of 35,000 feet during a July 1943 test flight were considered to be only moderately better than the original prototype.[131]

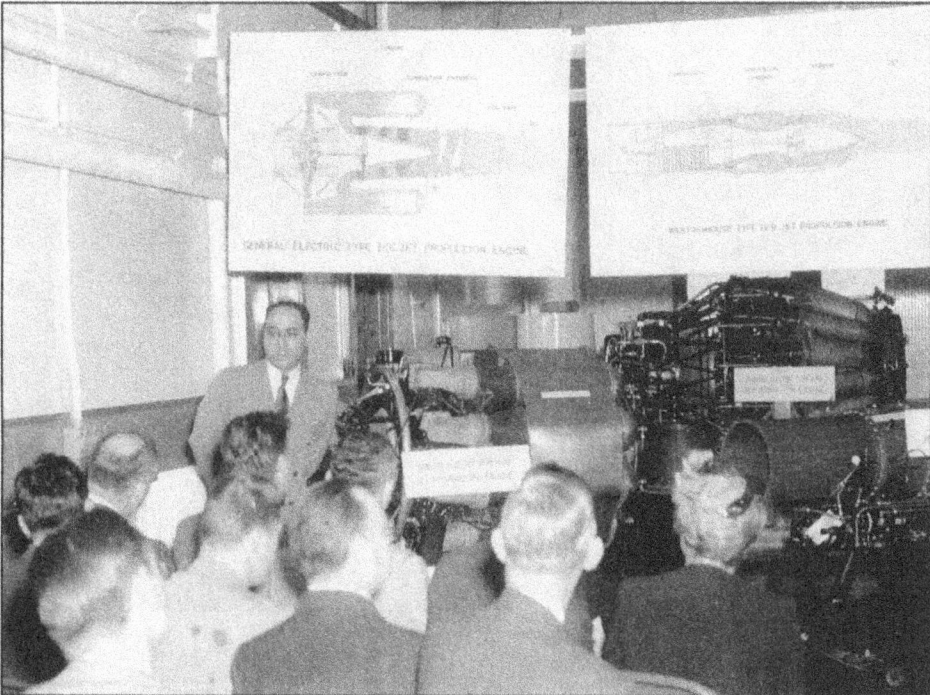

Image 46: Benjamin Pinkel uses a GE I-16 engine to explain thrust augmentation studies to members of the Aviation Writer's Association touring the AERL. The GE I-16 and I-40 and Westinghouse 19B engines on display were tested in the AWT during the war years. Pinkel had worked on the Jeep engine project at Langley before being transferred to Cleveland in December 1942. (NASA C-1945-10634)

The Turbojet Takes Precedence

By the end of the year it had been decided that the GE I-16, not the long-waiting Wright R-3350, would be the first engine investigated in the NACA's new AWT. Colonel Keirn summoned Silverstein, the head of the AWT, to GE's West Lynn facility to discuss testing the I-16 in altitude conditions. When asked later about the trip, Silverstein replied, "We had the only altitude tunnel. That's the reason I went." Silverstein had been aware of the Jeep project, but the meeting with Keirn was his first real exposure to the turbojet. After viewing the work at West Lynn, arrangements were made "to get the engine going."[132]

Although the AWT was not designed for jet engines, Silverstein leapt at the chance to examine the new technology in his brand new wind tunnel. One of the I-16's most troublesome problems was the uneven airflow through

Image 47: The AWT's first test: the Bell YP-59A with its two GE I-16 turbojet engines. Improvements from the AWT tests included a boundary-layer removal duct, which decreased the fuselage's boundary layer by 60 percent; a new nacelle inlet, which in combination with the boundary-layer removal duct resulted in an additional 16-percent average pressure recovery near the compressor inlets; and new engine cooling seals, which reduced the nacelle's cooling airflow by 75 percent.[133] (NASA C–1944–04825)

Image 48: Bell YP-59A, a production version of the prototype XP-59A, was flown from the Bell plant in Buffalo, New York, to the AERL by Bob Stanley. Stanley piloted the first successful flight of the XP-59A at Muroc Army Air Field (now Edwards Air Force Base) on 1 October 1942. The secret AWT tests led by Merritt Preston improved the performance of the engine, but the enhancements could not overcome the engine's many design flaws. (NASA C–1944–04314)

its intakes. Silverstein had already investigated engine airflow in Langley's Full-Scale Tunnel and felt confident that studies in the AWT would result in increased thrust for the I-16.[134]

After the logistics were worked out, Bell pilot Bob Stanley flew a YP-59A to Cleveland for the tests.[135] Harold Friedman remembered the aircraft's arrival, "[Colonel Page, the U.S. Army Air Force liaison] was all dressed up in his air force uniform and was carrying a sidearm like he was going to protect it."[136] The wing tips and tail were cut from the aircraft so that the entire fuselage and engines would fit into the AWT's test section. The tests began on 4 February 1944.

The study, led by Merritt Preston, first analyzed the engines in their original configuration and then implemented a boundary layer removal duct, a new nacelle inlet, and new cooling seals. Tests of the modified version showed that the improved distribution of airflow increased the I-16's performance by 25 percent.[137] Despite the improved speed, the aircraft was not stable enough to be used in combat, and the design was soon abandoned. GE created 241 of the engines for the U.S. Army Air Force, but only 20 production Airacomets were built by Bell. These were used primarily for pilot training.[138]

Image 49: The Bell YP-59A Airacomet with nacelle cover removed to reveal one of its two GE I-16 jet engines. GE's West Lynn plant had developed the turbocharger, which increased the capabilities of piston engines and allowed the first over-weather flight in July 1937. Because of the demand for the turbochargers, the West Lynn plant expanded its facilities and began working closely with the air corps. This relationship led to its selection to build the nation's first jet engine, the I-A. Turbocharger guru Sanford Moss was called out of retirement in 1943 to assist with the next generation of centrifugal jets—the I-14, I-16, and I-20.[139] (NASA C–1945–10686)

Image 50: Mechanics attend a class on jet propulsion. The introduction of turbojets during the war required crash studies for mechanics and engineers alike. In this photograph, Gesa Major demonstrates a centrifugal compressor used on the Whittle-based engines. (NASA C–1945–13464)

"Too Much Damn Work"

The nation made countless sacrifices during the war, and the AERL was no exception. The January 1943 Overtime Act resulted in a six-day, 48-hour work-week without overtime pay for the duration of the war. Those earning under $2,900 were given a 22 percent salary increase, but the 5 percent "Victory Tax" remained in place.[140] The AWT crew worked around the clock to keep the facility going.

There were many tests in its queue when the AWT became operational in February 1944. As soon as one investigation was complete, the engine was quickly removed and replaced with another. For example, mechanics were prepping the Douglas XTB2D and Westinghouse 19B engines in the shop as the B-29's R-3350 engine was being tested. After the R-3350 failed at 3 a.m. on 14 September 1944, it was quickly removed from the test section. By 16 September, a 19B had taken its place. After adding a lengthy list of tasks completed on 5 September 1944 to the AWT logbook, mechanical supervisor, Austin Reader, noted, "Too much damn work."[141]

Image 51: The AWT seen during its overnight operation. The tunnel was usually run at night when the electric company had sufficient electricity for the massive power loads. (NASA C–1945–09513)

Operating the AWT was a demanding job, even without the pressure of the war. For the mechanics, the work was often dirty and tedious. For the engineers, the days were long and stressful. Generally the first and second shifts set up or broke down the test articles. The tests were run by the third shift at night when electricity was available. Engineers would often have to work all day, then operate the tunnel and run their test overnight.

NASA retiree Bob Walker described it: "First and second shift [were] designed to get the work done for the day, all the squawks on your squawk sheet, just like you'd have an aircraft operation someplace. The engineers would put down the work they wanted done and things you'd burned up the night before or whatever. So, you'd get it all back together on the first and second shift, then [the overnight crew] would come in and crank it up at night and run it."[142]

Image 52: Crowded conditions inside the Fabrication Shop. The lab was under pressure to meet the military's demand for wartime testing. This resulted in 48-hour weeks and three shifts per day. The AWT's late entry into the war left no time to spare. Those who were frequently absent or arrived late were transferred or discharged.[143] (NASA C-1945-10396)

Harold Friedman added, "We were running the Altitude Wind Tunnel from 10:00 at night until 4:00 in the morning because that's when the electricity was available. When that was finished, we'd go home and go to bed and get up and come back to work."[144]

"Because there were fewer and fewer engineers who wanted third shift after working all day at their desks," recalled former technician Howard Wine, "they came up with a theory that, well, maybe the technicians, the crew chiefs, could operate the facility." The operation of the tunnel was a sophisticated process that required communication with the electric company and the support buildings. According to Wine, the crew chiefs were trained to operate the tunnels, even though some of the mechanics had a better understanding of the system.[145]

Image 53: Lockheed YP-80A with its GE I-40 engines being tested in the AWT. One of the modifications required to test jet engines in the AWT was the ducting of conditioned airflow directly to the engine inlets. This method effectively reduced the tunnel size but allowed greater speeds and altitude conditions for the engine. The direct-connect method was used for all future turbojet and ramjet tests in the AWT. (NASA C-1945-09446)

"It Was So Simple"

In May 1944, a wing section with the right inboard nacelle and its 18-cylinder R-3350 engine was finally installed in the AWT. Components of the R-3350 had been investigated the previous fall in the Engine Research Building, resulting in an improved fuel-injection system, but the AWT would be needed to study the entire engine under the normal high-altitude environment, particularly the airflow around the engine.

Abe Silverstein played a key role in the studies. Not long before his transfer from NACA Langley, he began studying engine cooling in the Full-Scale Tunnel. He used his acquired aerodynamics knowledge to improve the flow of air through the engines.[146] The AERL engineers found that there was no

Image 54: The AWT was used to study the engine cooling problems for the Boeing B-29 Superfortress for the Pacific phase of World War II. The B-29's right inboard nacelle and wing section were installed in the AWT test section from May to September 1944. As part of this series of tests, several different flap designs were studied on the Wright R-3350 engine's 43-inch-diameter cowl inlet. The logbook notes that the day shift "washed model for pictures" prior to this midday 4 July photograph.[147] (NASA C–1944–05554)

Image 55: The massive Wright R-3350 18-cylinder piston engine is prepared for testing in the AWT. In 1935, Curtiss-Wright Corporation began developing bigger and more powerful aircraft engines that were based on the principles of their existing Cyclone engine. One of the new engines was the 2,200-horsepower R-3350, which was plagued by problems from the beginning. Although first run in May 1937, it was not successfully flown until 21 September 1941. When proposals were drawn for the B-29 bomber in 1940, the design included the use of four of the unproven R-3350s. (NASA C-1944-04488)

cooling airflow where the exhaust was ejected, which was the hottest section of the engine. After a relatively short period of analysis, they found that by elongating fins within the engine the airflow would be directed around the cylinders to the problem areas. This resulted in a 50°F reduction of the operating temperature. Silverstein, who was gaining more and more experience with engine cooling, later explained, "It was so simple."[148]

The AWT also was used to study the R-3350's cowl inlets, particularly the flap design. Cooling-air pressure levels, distribution, and drag were analyzed for a variety of cowl flap configurations. The researchers found that sliding flaps required 60 to 80 less horsepower than did the original chord flaps. This would produce an extra 190 mph at an altitude of 15,000 feet.[149]

The design of the R-3350 still had problems, but the AERL enhancements could get the bomber through the war. The 18 percent increase in fuel efficiency could broaden the B-29's flight range or increase its armament capabilities.[150] This improvement was calculated to be either an altitude increase of 10,000 feet, a gross-weight increase of 10,000 pounds at sea level, or a gross weight increase of 35,000 pounds above 10,000 feet.[151]

The final run of the R-3350 in the AWT had an inauspicious conclusion on 15 September 1944. "At approx[imately] 3:02AM this morning the B-29 threw #14 cylinder of the engine during operation the cyl[inder], passed thr[ough] the cowling[,] hit tunnel under wing, hit base of exh[a]ust inlet & hit half way up turning vanes putting large dent in vane. The piston was found at the second vane. Most of piston was stopped by screen at this vane. The tunnel was locked up awaiting inspection of U.S. Army. Took pictures[,] then [Royce] Moore & Silverstein ordered B29 removed from tunnel."[152] Nonetheless, the studies had been successful, and some felt that just 10 days of R-3350 testing had paid for the NACA's investment in the AWT.[153]

Forsaken

Harold Friedman and others from the AERL flew to Boeing headquarters in Renton, Washington, to present the findings from the AERL tests. The group remained in Washington for six weeks while Boeing verified the AERL conclusions with their own tests. Friedman claimed that the suggested modifications were not implemented. He surmised that it was because the fuel-injection system would have increased fuel consumption.[154] It appears, however, that the findings arrived just a little too late for the modifications to be used for the war.

In late August 1944, just as the AWT tests were wrapping up, Major General Curtis LeMay was placed in charge of the XXIst Bomber Command, which controlled all B-29 aircraft in the Pacific. The first phase of the B-29 bombing campaign, beginning three months later in November 1944, consisted of high-altitude precision strikes on specific targets during daylight.

In addition to the engine difficulties, the winds at high altitudes were causing navigation problems. General Thomas Power, who led the B-29 missions, and LeMay decided to switch their strategy to low-level missions to avoid the weather problems and keep the engines from burning up. There would be no need to bolster the R-3350s for high-altitude flight, and the reduction of engine overhauls would provide additional bombers for the sorties. In addition, the aircraft could carry a larger quantity of bombs, hit their targets more accurately, and use less fuel.[155]

The second phase of bombing was at low-altitude at nighttime. It began during the night of 9 March 1945. General Power led three-hundred forty-six B-29s on an assault of Tokyo using napalmlike incendiary bombs. The attack killed 84,000 Japanese and destroyed 16 square miles of the city. Postwar studies indicated that the fires on this night produced more heat than the Dresden bombing or either of the subsequent atomic bombs.[156] The devastating low-level nighttime fire-bomb raids and medium-altitude daylight attacks continued through the end of the war.[157]

Image 56: B-29 Superfortress in the AERL hangar for a postwar open house event with its Wright R-3350 engine on display. The AERL had remedied the engine's cooling problems, but many of the modifications would not be implemented until after the war. (NASA C-1945-10587)

It is difficult to say for certain, but if the R-3350 testing in the AWT had not been delayed for several months, the U.S. Army Air Force may have been able to accomplish its goals by using the B-29's high-altitude capabilities to bomb Japanese industrial and transportation targets, and the incendiary bombing—which resulted in so many casualties—may have been averted. After the war the R-3350 went on to a successful career, powering both civilian and military aircraft using the baffling and fuel-injection modifications made at the AERL. The precedence given to the turbojet would pay off for the AWT, though. By the time of the March raid on Tokyo, the AWT had investigated three jet engines, including the nation's first successful jet fighter, the Lockheed Shooting Star.

The Americans Get It Right

Although the AWT investigations significantly increased the I-16's thrust, the military needed an engine twice as powerful. The Messerschmitt Me-262 had begun making its mark in Europe, and the air force sought a new 500-mph fighter.[158] In June 1943, GE engineers began work on a 4,200-pound-thrust engine. The new I-40 engine was first tested at West Lynn, Massachusetts, the following January, just as the AWT was preparing to test the I-16s.[159]

Image 57: Lockheed's YP-80A Shooting Star on display in the AWT shop. The Shooting Star was the first jet aircraft manufactured in the United States and the first U.S. Air Force aircraft to fly faster than 500 mph. (NASA C–1945–10600)

After the disappointment of Bell's XP-59A, in June 1943 the air force had tapped Lockheed to design a new jet fighter to incorporate the I-40s. Lockheed's drawings for the XP-80 Shooting Star were approved on 16 October, and the first aircraft, which used Halford H-1 engines, was completed 150 days later. The Shooting Star's initial flight took place in January 1944 as construction of the AWT was being finalized.[160]

The flight was a success, but almost immediately the airframe was modified for use with the I-40. Two new XP-80As were produced and flight tested during the summer of 1944. On the basis of their performance, Lockheed produced the first eighteen YP-80A Shooting Star production aircraft in September. The air force dispatched two to Britain and two to Italy to try to neutralize the Me-262's successes. The Shooting Star continued to experience operational problems, though, and crashes resulted in the deaths of several pilots.[161]

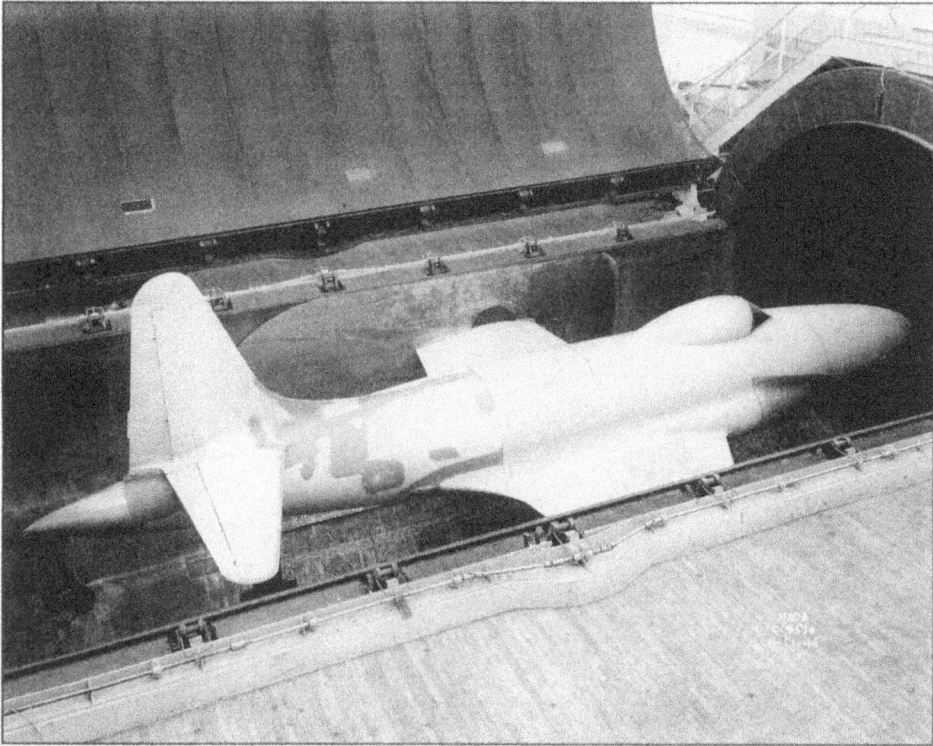

Image 58: GE I-40 engines for Lockheed's YP-80A Shooting Star were tested in the AWT from March to May 1945. The tunnel's 20-foot-diameter test section allowed the entire fuselage to be installed. (NASA C–1945–09576)

The I-40 engine underwent a thorough analysis in the AWT during the spring of 1945. Like the Airacomet, the entire YP-80A fuselage was installed in the test section. The tunnel airflow was directly connected to the inlets in order to increase the altitude and speed of the air as it entered the engine. One of the primary areas of research was the engine's thrust performance at altitudes up to 50,000 feet. An attempt to forecast altitude thrust levels based on sea-level measurements was successful, and a curve was created to predict the I-40's thrust at all altitudes.[162]

Follow-up studies on a Lockheed TP80S, a Shooting Star modified to accommodate a second pilot, revealed that the I-40's turbine and compressor efficiency and fuel consumption were not affected by altitude but that combustion efficiency and thrust diminished as altitude increased.[163] After analyzing different tailpipes, researchers found that a short nozzle with a uniform diameter tailpipe was most efficient.[164]

The resulting P-80 fighter and the I-40 engines would be great successes, just not in time for the war effort. Thousands of the I-40 engines were eventually built, including 300 during the last year of the war. As a result of the I-40, for the first time British engineers were traveling to the United States to study a jet engine.[165] In 1947 the P-80 set the world's speed record with a 620-mph flight at the Cleveland Air Races. The second generation, the F-80, was a vital weapon in the Korean War.[166]

Endnotes for Chapter 3

103. Ruth Baker, "Research Staff Meetings Inaugurated," Wing Tip (8 January 1943).

104. John Curatola, "No Quarter Give: The Change in the Strategic Bombing Application in the Pacific Theater During World War II" (Thesis, U.S. Army Command and General Staff College, 2002), 57.

105. Robert F. Dorr, *Superfortress Units of World War 2* (Osceola, WI: Osprey Publishing, 2002), p. 9.

106. Henry H. Arnold, "Transcript of General H.H. Arnold's Speech to AERL," 9 November 1944, File 4849, Historical Collection, NASA Headquarters, Washington, DC.

107. "Utilization of Wind Tunnels From January 1939 to June 1945," File 11059, Historical Collection, NASA Headquarters, Washington, DC.

108. Kimble McCutheon, "OX-5s to Turbo-Compounds: A Brief Overview of Aircraft Engine Development," 1999, p. 1.

109. Dorr, *Superfortress Units of WWII*.

110. Curatola, "No Quarter Give," pp. 59–60.

111. Dorr, *Superfortress Units of WWII*.

112. Curatola, "No Quarter Give," pp. 61–63.

113. McCutheon, "OX-5s to Turbo-Compounds," p. 10.

114. Interview with Frank Bechtel, conducted by Bob Arrighi, 19 July 2004, NASA Glenn History Collection, Cleveland, OH.

115. Dik Daso, "Origins of Air Power: Hap Arnold's Command Years and Aviation Technology, 1936–1945," *Aerospace Power Journal* (Fall 1997): 99 and 110.

116. Jerome Hunsaker to George Lewis, "Memorandum of Conference With General Echols Regarding Suggestion Contained in General Arnold's Letter of October 14, 1942, to Dr. Hunsaker, Relating to Engine Design Improvements Especially for Fighters," 23 January 1943, File 11059, Historical Collection, NASA Headquarters, Washington, DC.

117. T.A. Heppenheimer, "The Jet Plane Is Born," *American Heritage: Invention and Technology Magazine* I, no. 2 (Fall 1993).

118. James R. Hansen, *Engineer in Charge: A History of the Langley Aeronautical Laboratory, 1917–1958* (Washington, DC: NASA SP–4305, 1987), chap. 8.

119. Interview with Abe Silverstein, conducted by Virginia Dawson, 5 October 1984, NASA Glenn History Collection, Cleveland, OH.

120. Heppenheimer, "The Jet Plane Is Born."

121. "Jet Propulsion in England: Report of a Conference Attended by Officers of the Army and Navy and Representatives of American Industrial Organizations Visiting England During the Summer and Fall of 1943," 18 December 1943, File 11059, Historical Collection, NASA Headquarters, Washington, DC.

122. M. Neufeld and R. Lee, *Messerschmitt Me 262A-1a* (Washington, DC: National Air and Space Museum, Smithsonian Institution, 2 April 2001), *http://www.nasm.si.edu/research/aero/aircraft/me262.htm* (accessed 29 July 2009).

123. Silverstein interview, conducted by Virginia Dawson, 5 October 1984.

124. Hansen, *Engineer in Charge.*

125. Hansen, *Engineer in Charge.*

126. General Electric Company, *Seven Decades of Progress: A Heritage of Aircraft Turbine Technology* (Fallbrook, California: Aero Publishers, Inc., 1979), p. 48.

127. Jos Heyman and Andreas Parsch, "Duplications in U.S. Military Aircraft Designation Series," Designation-Systems.net (2004), *http://www.designation-systems.net/usmilav/duplications.html* (accessed 29 July 2009).

128. General Electric Company, *Seven Decades of Progress.*

129. David M. Carpenter, *Flame Powered: The Bell XP-59A Airacomet and the General Electric I-A Engine* (Jet Pioneers of America, 1992), pp. 38–39.

130. Henry H. Arnold to Jerome Hunsaker, 14 October 1942, File 11059, NASA Headquarters Historical Collection, Washington, DC.

131. Arnold to Hunsaker, "Letter of 14 October 1942."

132. Merritt Preston, Fred O. Black, and James M. Jagger, *Altitude-Wind-Tunnel Tests of Power-Plant Installation in Jet-Propelled Fighter* (Washington, DC: NACA MR–E5L17, 1946), p. 1.

133. General Electric Company, *Seven Decades of Progress,* p. 52.

134. Interview with Abe Silverstein, conducted by John Sloop, 29 May 1974.

135. Interview with Abe Silverstein, conducted by Walter T. Bonney, 21 October 1972 and 20 September 1973, NASA Glenn History Collection, Oral History Collection, Cleveland, OH.

136. Silverstein interview, conducted by Walter T. Bonney, 21 October 1972 and 20 September 1973.

137. Interview with Harold Friedman, conducted by Bob Arrighi, 2 November 2005, NASA Glenn History Collection, Oral History Collection, Cleveland, OH.

138. Preston, "Altitude-Wind-Tunnel Tests."

139. General Electric Company, *Seven Decades of Progress,* p. 52.

140. General Electric Company, *Seven Decades of Progress,* p. 52.

141. "It's Now Six-Day, 48-Hour Week," *Wing Tips* (22 January 1943).

142. "Day Shift," *AWT Logbook Record: June 26, 1944 to February 28, 1945* (Cleveland, OH: NASA Glenn History Collection, 5 September 1944).

143. Interview with Howard Wine, conducted by Bob Arrighi, 4 September 2005, NASA Glenn History Collection, Cleveland, OH.

144. Interview with Bob Walker, conducted by Bob Arrighi, 2 August 2005, NASA Glenn History Collection, Cleveland, OH.

145. Friedman Interview, conducted by Bob Arrighi, 2 November 2005.

146. Silverstein interview, conducted by Walter T. Bonney, 21 October 1972 and 20 September 1973.

147. "Day Shift," *AWT Log Record: June 26, 1944 to February 28, 1945.*

148. Silverstein interview, conducted by Virginia Dawson, 5 October 1984.

149. DeMarquis Wyatt and William Conrad, *An Investigation of Cowl-Flap and Cowl-Outlet Designs for the B–29 Power Plant Installation* (Washington, DC: NACA Wartime Report E–205, 1946), p. 27.

150. "Center Marks Thirty Years Progress," *Lewis News* (16 July 1971).

151. Frank Marble, Mahlon Miller, and Barton Bell, "Analysis of Cooling Limitations and Effect of Engine Cooling Improvements on Level Flight Cruising Performance of Four Engine Heavy Bomber" (Washington, DC: NACA Report 860, 1948).

152. "Day Shift," *AWT Log Record: June 26, 1944 to February 28, 1945.*

153. L.L. Lewis memo, "Carrier in World War II Wind Tunnel Air Conditioning," United Technologies Archives, as referenced in Virginia P. Dawson, *Engines and Innovation: Lewis Laboratory and American Propulsion Technology.* (Washington, DC: NASA SP–4306, 1991), chap. 2.

154. Friedman interview, conducted by Bob Arrighi, 2 November 2005.

155. Dorr, *Superfortress Units of WWII,* pp. 23 and 36.

156. Dorr, *Superfortress Units of WWII,* pp. 2 and 38.

157. Curatola, "No Quarter Give," 58.

158. General Electric Company, *Seven Decades of Progress,* p. 52.

159. General Electric Company, *Seven Decades of Progress,* p. 53.

160. Laurence K. Loftkin, Jr., *Quest for Performance: The Evolution of Modern Aircraft* (Washington, DC: NASA SP–468, 1985), chap. 11, *http://www.hq.nasa.gov/pao/History/SP-468/ch11-2.htm* (accessed 30 July 2009).

161. Marcelle Size Knaack, *Encyclopedia of U.S. Air Force Aircraft and Missile System,* Volume I (Washington, DC: Office of Air Force History, 1978), pp. 1–3.

162. Abe Silverstein, "Altitude Wind Tunnel Investigations of Jet-Propulsion Engines," General Electric Gas Turbine Conference (31 May 1945), pp. 7–10.

163. Stanley Gendler and William Koffel, *Investigation of the I-40 Jet Propulsion Engine in the Cleveland Altitude Wind Tunnel I—Performance and Windmilling Drag Characteristics* (Washington, DC: NACA RM–E8G02, 1948), p. 16.

164. Richard Krebs and Frederick Foshag, *Investigation of the I-40 Jet Propulsion Engine in the Cleveland Altitude Wind Tunnel III—Analysis of Turbine Performance and Effect of Tail-Pipe Design on Engine Performance* (Washington, DC: NACA RM–E8G02b, 1948), p. 3.

165. General Electric Company, *Seven Decades of Progress,* p. 55.

166. Knaack, *Encyclopedia of U.S. Air Force Aircraft,* p. 1–3.

Image 59: A researcher examines the stator blades on a Westinghouse 24C axial-flow turbojet engine in the shop area. Axial-flow engines, with multiple stages or rows of compressor blades as shown here, were studied extensively in the Altitude Wind Tunnel. (NASA C-1950-26086)

Chapter 4

Age of the Axial Flow | Alternate Wartime Mission (1944–1945)

The new turbojet reorganization at the NACA's Aircraft Engine Research Laboratory (AERL), known as "the Big Switch," would not officially take place until the end of the war. The emphasis on jet engines, however, unofficially began two years earlier with the acquisition of the General Electric (GE) centrifugal engines for studies in the Jet Propulsion Static Lab and then the Altitude Wind Tunnel (AWT). The sudden emergence of the turbojet affected the AERL far more than the NACA's two aerodynamics-based laboratories. The staff had to quickly learn the new technology and modify their test facilities to accommodate more powerful engines.

One of the most significant problems with the early turbojets was combustion at high altitudes. The AWT was the nation's only facility in which the combustion and performance characteristics of a turbojet could be studied under altitude conditions.[167] During the final year of the war, the AWT was used almost exclusively to address this problem. In addition to testing the Whittle-based series of centrifugal engines, the tunnel was used to study and improve almost every early model of the axial-flow compressor engine. The axial-flow engine powered the successful Messerschmitt Me-262 and would prove to be the enduring version of the turbojet.

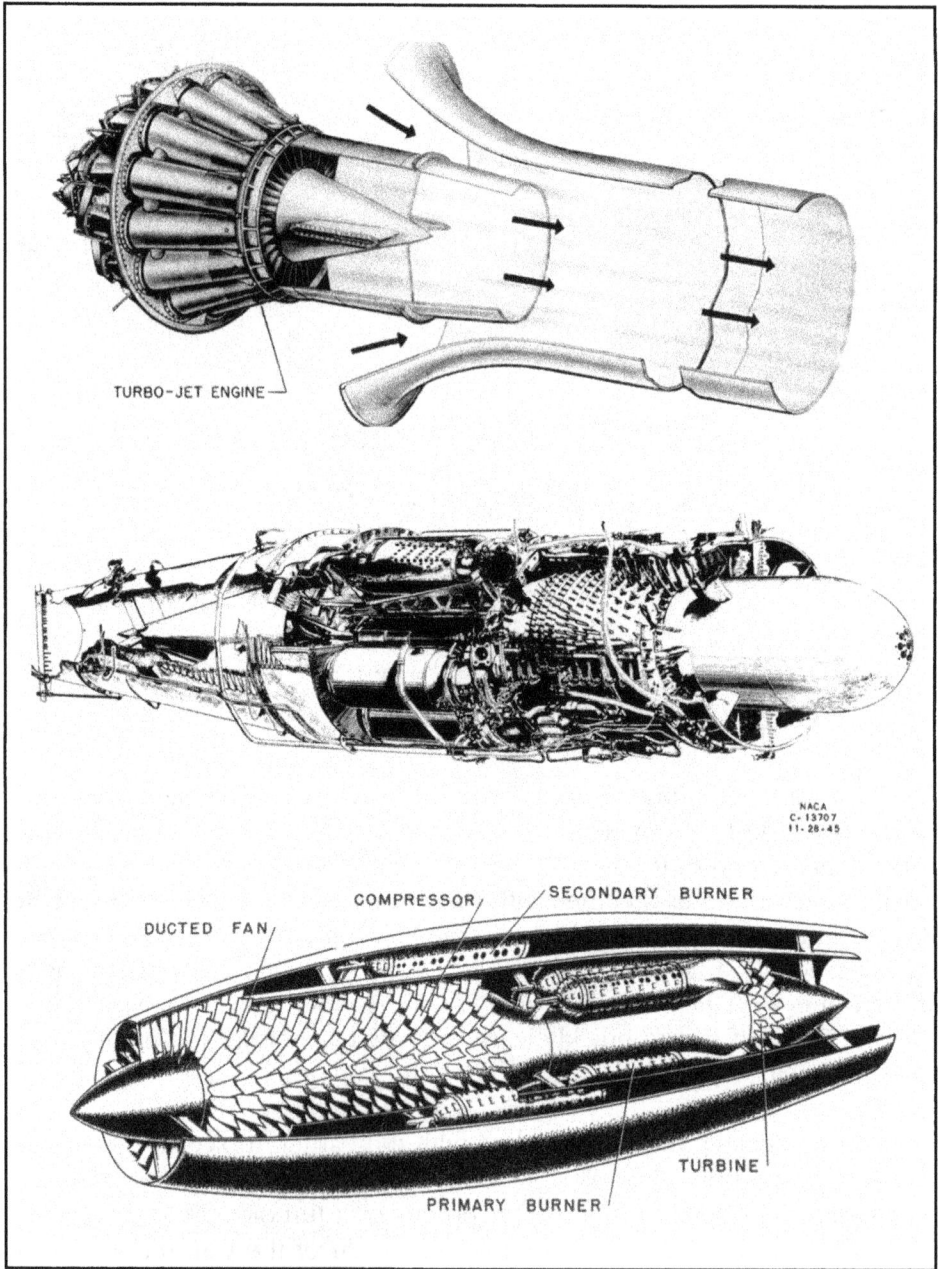

TURBO-JET ENGINE

NACA
C-13707
11-28-45

COMPRESSOR   SECONDARY BURNER

DUCTED FAN

TURBINE

PRIMARY BURNER

Image 60: Cutaway drawings. Top: Centrifugal compressor engine. Center: Axial-flow engine. Bottom: Ducted-fan engine. Growth of the centrifugal engine was limited by the outward expansion of the single compressor. At a certain point, the tradeoff between power and size would be inefficient. Axial-flow engines could gain additional power by adding sets of compressor blades in a line. The length of the engine would increase but not the diameter. (NASA C-1946–15565, C-1945–13707, and C-1946–15563)

Steam Turbine Experts Heed the Call-Up

Although the military had considered as many as 15 different jet engines from a number of different companies between 1942 and 1945, the two experienced steam turbine manufacturers—Westinghouse, and GE's Schenectady, New York, plant—produced the most significant results. William Durand's Special Committee on Jet Propulsion included Westinghouse's L. W. Chubb, Schenectady's Alan Howard, and Allis-Chalmers's R. C. Allen. The largest aircraft engine manufacturers, Pratt & Whitney and Wright Aeronautical, were omitted so that they could focus their efforts on producing piston engines for the war.[168] Westinghouse contracted with the navy's Bureau of Aeronautics and GE with the U.S. Army Air Corps. During the 1940s, both of the companies created three increasingly powerful generations of axial-flow turbojets.

In July 1941, months before GE's West Lynn group began recreating Whittle's engine, Durand's committee asked three engine companies to each design a jet engine for possible military production. Allis-Chalmers proposed a ducted fan that remained mired in the design process, GE suggested the TG-100

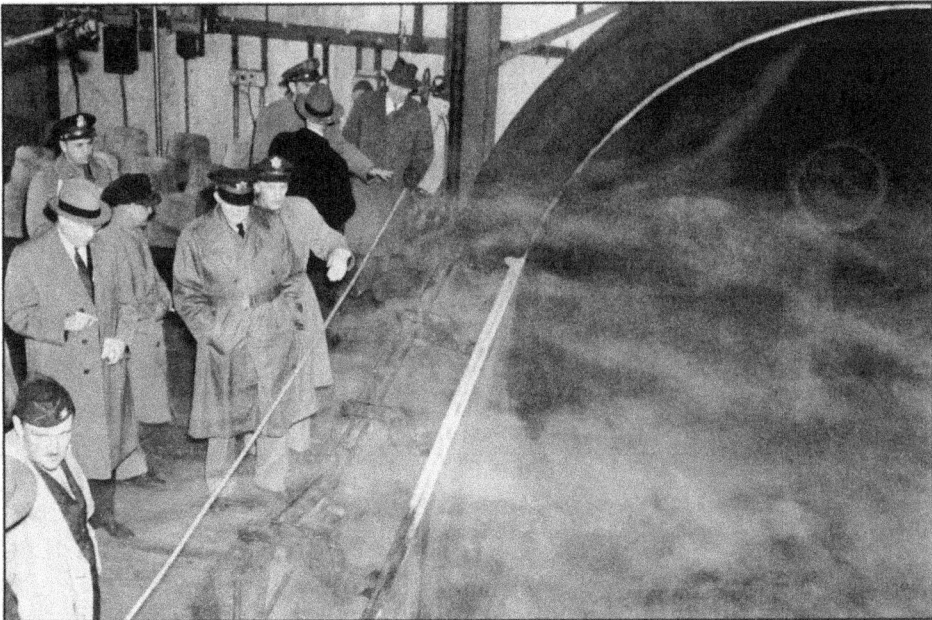

Image 61: NACA leaders accompany General Henry Arnold on his tour of the AWT on 9 November 1944. Here the group is on the tunnel's viewing platform peering into the 20-foot-diameter test section. The refrigerated air in the tunnel is vaporizing as it mixes with the warmer exterior air. Arnold pushed the development of jet engines during the war. This included the testing of nearly every early model in the AWT. (NASA C–1944–07499)

turboprop, and Westinghouse offered the 19A turbojet. All were based on the axial-flow compressor model.[169] The initial designs quickly led to more powerful incarnations. The AWT would conduct extensive studies on almost every one of the early axial-flow engines. Each of these studies included an analysis of general operating characteristics over a range of speeds and altitudes, as well as analysis of specific performance enhancements such as tailpipe burning, windmilling, and high-altitude flameouts.

Westinghouse Designs First U.S. Turbojet

In November 1941 the navy requested that Westinghouse put together a proposal to design a turbojet engine capable of reaching 500 mph. The day after the Pearl Harbor attack, despite not having a contract, Westinghouse began work on what would be the 19A engine. Official work began the following August, and the 19A was first run on 19 March 1943. Following a 100-hour endurance test on 5 July, the 19A became the first operational jet engine designed in the United States.[170] When it was used on 21 January 1944 as a jet-assisted-takeoff-like booster on a Chance-Vought FG-1 Corsair, the 19A became the first and only of the original U.S. turbojets to be flown during the war in Europe.

Image 62: Westinghouse 19B six-stage axial-flow turbojet in the AWT. The 1,400-pound-thrust 19B improved on Westinghouse's original 1,100-pound-thrust 19A engine. Instrument rakes can be seen in the engine's inlet. The engines were relatively small, with a length of 8-feet 8.5 inches, with a maximum diameter of 20.75 inches, and a weight of 825 pounds.[171] (NASA C-1944-06735)

In March 1943, just over a week before the first run of the 19A, Westinghouse agreed to create an improved six-stage version, the 19B. Unlike its predecessor, the 19B could serve as either a booster or primary propulsion unit. The engine underwent its first test run a year later in March 1944. Almost immediately the navy agreed to Westinghouse's proposal for the even larger 10-stage, 1,600-pound-thrust 19XB prototype.[172] By July the navy had contracted with the NACA for the testing of both engines in the AWT.

The AWT investigations began on 9 September, one week after the 19B underwent its first flight test. The AWT runs revealed the superiority of the previously untested 19XB over the 19B. The 19B engines failed to restart consistently and suffered combustion blowouts above 17,000 feet. The 19XB, however, performed well and restarted routinely at twice that altitude.[173] Two months later on 26 January 1945, two 19Bs powered a McDonnell XFD-1 Phantom, the U.S. Navy's first fighter jet, on its initial flight. Following its exceptional performance in the AWT, the 19XB engines soon replaced the 19Bs in the Phantom.[174]

Image 63: The Westinghouse 19XB seen from the side in the AWT. Two 19XBs arrived on 6 November 1944 and were promptly installed in the tunnel. The 19XB-1 and 19XB-2B produced 1,400 and 1,600 pounds of thrust, respectively. The AWT tests revealed the superiority of the 19XB over the 19B and led to its success. (NASA C-1944-07564)

General Electric Takes on the Axial-Flow Engine

When asked to participate in the Special Committee on Jet Propulsion in March 1942, GE was still trying to build the Whittle engine at its West Lynn, Massachusetts, plant. Another GE group in Schenectady, New York, however, had been working on the TG-100 axial-flow turboprop engine for several years. Turboprops could move a large volume of air and thus required less engine speed. The turboprop was progressing slowly, so in May 1943 the army requested the development of an axial-flow turbojet, the TG-180. It was accepted that the TG-180 would not be ready in time for the war effort, but its long-term potential was considered to be more promising than that of the Whittle engines.[175]

Like the turboprop, the development of the TG-180 was sluggish. Although the engine was bench tested in April 1944, it was not flight tested until February 1946.[176] During the interim, the engine was brought to the AERL

Image 64: The McDonnell Phantom XFD-1 became the U.S. Navy's first completely jet-propelled aircraft. Its development began in August 1943. The Phantom was originally designed to use two Westinghouse 19B jet engines. After successful tests in the AWT, these were replaced by 19XBs. The 19XBs were also used for the experimental Douglas XB-42A Mixmaster and the Northrop XP-79 "flying wing" aircraft. (U.S. Navy/Department of Defense)

for the first of four rounds of testing in the AWT. The studies, which would continue intermittently into 1948, subjected the engine to an array of tests. Modifications were made that steadily improved the TG-180's performance, including the first successful use of an afterburner.

To GE's chagrin, the army contracted with Allison to manufacture the TG-180s for use on the Bell X-5, Republic F-84 Thunderjet, and Northrop F-89 Scorpion. Although the TG-180 was not the breakthrough engine that the military had hoped for, it did power the Douglas D-558-I Skystreak to a world speed record on 20 August 1947.[177]

Image 65: Douglas D-558-I Skystreak powered by the TG-180 engine. On 20 August 1947 U.S. Navy Commander T. Caldwell flew the Skystreak to a new world's speed record of 640.7 mph. A second D-558-I flown by U.S. Marine Pilot Major Marion Carl soon bettered the record by 10 mph. Unlike the X-planes, the Skystreaks took off from the ground under their own power and had straight wings and tails.[178] (NASA E-713)

Image 66: Eleven-stage axial-flow GE TG-180 installed in the AWT test section. The initial tests focused on the benefits of tailpipe burners, or afterburners, to improve thrust. One area of research determined the performance of a 29-inch-diameter tailpipe burner over a range of altitude conditions using several different flameholders and fuel systems. AERL researchers determined that the optimal design was a three-stage flameholder with its largest stage upstream.[179] (NASA C-1945-08687)

"It Was No Accident"

Once the decision was made to proceed, the development of turbojet engines was given precedence and heavy financial support by the military. Numerous sea-level facilities for testing piston engine endurance existed, and others were created specifically for the turbojet. One of the difficulties in developing turbojets, however, was the lack of altitude test facilities capable of handling the jet engines. The army's Wright Field and the navy's Aeronautical Engine Laboratory had a number of test stands, including altitude simulators, but the range of conditions and the size of the engines were restricted. The same was true for the altitude stands at the National Bureau of Standards and GE. Flight tests were complicated because it was extremely difficult to fly a test aircraft with elaborate instrumentation installed on the jet engines. The AWT was the only facility the nation had to study the performance of large turbojets at altitude conditions.[180]

Image 67: Left to right: Roger Weining and Bob Godman operate an engine from the AWT's control room. The tunnel's ability to repeatedly study a full-size engine under flight conditions and engine parameters of their choosing allowed the researchers to make modifications and test them quickly without the expense and uncontrolled variables of flight testing. (NASA C-1945-10365)

This ability permitted the AERL engineers to try different modifications to improve performance without redesigning the entire engine. Former AERL engineer Bruce Lundin recalled, "We had the facilities and the opportunity to create new kinds of compressors, to create afterburners, to create variable-area nozzles, to create high-altitude combustors and all this stuff...let's try this altitude, and let's see if this will work. If it doesn't work, we'll fix it so it does. And then we can go to Boeing or Lockheed or General Dynamics or some of these places, show them what we've done and they can put the marketing skills into that, and the production skills, and the reliability, and the maintenance skills, and sell them. And that's why American aviation has led the world. It was no accident. It was this partnership where you each contribute to the building of the whole."[181]

During typical AWT investigations, the operational and performance data for the entire engine were determined at various altitude flight conditions, as well as that of the various engine components. The operating range of the

Image 68: Test engineers lower an NACA inlet duct into the 20-foot-diameter test section for installation on a Westinghouse J40 engine. Engines could easily be tested in flight conditions with modified components, alternate throttling methods, or different fuels. (NASA C-1951-28464)

engine was established at various altitudes, and it was determined whether the operating range was limited by high turbine temperature, faulty combustion, or other factors.[182] Once the range was established, the AWT researchers sought to make general improvements that could be used on any jet engine. Three of the most significant advances dealt with restarting at high altitudes, windmilling, and the afterburner.

Early versions of the turbojet suffered from combustion blowouts and were difficult to restart as the airspeed or altitude increased. As the mixture of fuel and air became colder and less dense at higher altitudes, ignition problems increased. The AWT was used to determine the maximum altitude at which a particular engine could be started. AWT testing of the GE TG-180 in 1945 showed that relatively small modifications such as spark plug adjustments resulted in considerable improvements. The engine started consistently at an altitude of 40,000 feet if the engine was properly throttled.[183]

Windmilling, or the rotation of compressor blades during flight while the engine is inoperative, causes serious aerodynamic drag on an aircraft. Reciprocating engines combat this by feathering the engines. In jet engines, the increased airflow from windmilling alters fuel-spray patterns, hampers ignition, and sharply increases the drag. AWT studies of both centrifugal and axial-flow engines revealed that windmilling was directly linked to airspeed but had no relation to altitude. At speeds of 600 mph the resulting drag exceeded the normal engine thrust for both the I-40 and TG-180.[184] Inlet closures and compressor brakes would be used in ensuing generations of jet engines to prevent this phenomenon.

The Afterburner Comes to Life

The advantages of the afterburner were agreed on almost immediately after the introduction of turbojet engines. The concept was not invented by AERL researchers, but they were the first to make it operational. Afterburners operate by heating the engine's exhaust before it fully expands. This produces additional thrust without increasing temperatures or stresses on the engines.[185]

The AWT was used extensively throughout the mid-1940s to study various afterburner configurations for several different engines. Each component— the tailpipe, flameholder, combustion chambers, and variable-area nozzle— was studied independently over a variety of altitudes and speeds. The researchers would have to weigh the tradeoffs between contradicting elements such as the need for cooling systems and the resulting extra weight.[186]

Abe Silverstein claimed that a 29-inch-diameter tailpipe installed in 1945 on a TG-180 in the AWT was the first operating afterburner. In a 1974 interview, Silverstein explained to John Sloop, "I recall very clearly the first night we ran that afterburner. I was sitting by one of [the AWT test section viewing

Image 69: Republic F-94 interceptor tests its afterburner prior to takeoff during the Korean War. The AWT was used to systematically study afterburner configurations on several early jet engines. Desired elements included maximum thrust, maximum operable range, high combustion efficiency, minimum weight, minimum size, low internal pressure losses, adequate cooling, and good control.[187] (27 July 1953) (National Archives)

Image 70: A technician adjusts the large afterburner on a GE TG-180 engine in the AWT test section. The 29-inch diameter of the burner resulted in a low-velocity, but highly efficient, thrust. The three main sections of tailpipe burners are the diffuser section, the combustion chamber, and nozzle section. The engine underwent thrust augmentation tests in the tunnel using a variety of tailpipe burners. (NASA C–1946–14938)

portals] when we turned that afterburner on for the first time. And of course they turned it on and immediately the thing combusted and ignited, and a flame 50 feet long, deep purple, came right out of the back end of that, a full flame, with a noise that was unbelievable. You wouldn't believe it." He continued, "We designed it for low velocity, a big area, and we got a very efficient afterburner, which was a good place to start." After the initial studies, the engineers began reducing the diameter to increase the velocity. Silverstein claimed that it took 10 years for NACA researchers to merge the more powerful smaller-diameter designs with the combustion efficiency achieved on that first run.[188]

The AERL worked closely with engine companies in the mid-1940s on improving the afterburner.[189] Although NACA researchers analyzed the burners for specific engines, they were able to accumulate a wide array of basic data that was applicable to most engines. By 1946 the lab had published data on the operation of afterburners, and the AWT would be used the following year to study a variety of components. These included the burner inlet diffusers, fuel-injection systems, flameholders, combustion space, combustion instability, starting performance, diluents, and burner shell cooling.[190] Within a decade, afterburners were incorporated into nearly all turbojet engines. In addition, their role had expanded from brief thrust augmentation to a key element of supersonic flight.[191]

Controlling the Thrust

Afterburner research produced several related innovations, including the cooling liner, V-gutter flameholder, fuel-spray bar system, and variable-area nozzle.[192] To combat the slow response time in early turbojets, AERL engineers developed an adjustable nozzle. This device was particularly important for takeoffs and landings. Bruce Lundin explained, "You can't have an afterburner if you don't have a variable-area exhaust nozzle. That was obvious from the beginning. The nozzle has to get larger when the afterburner's on; and when you turn it off, the nozzle's got to get smaller to keep the rest of engine operating properly."[193]

Image 71: A variable-area nozzle is installed on a GE TG-180 engine with an afterburner in the AWT. AERL created a clam-shell design because it was easier for the mechanics to build. The AERL studies led engine manufacturers to later develop an iris-type nozzle that was lighter and more efficient.[194] (NASA C-1945–8708)

After one of the AWT's overnight runs of an early afterburner, the crew informed Silverstein that the burner's shell was getting hot spots that could prove dangerous. Silverstein and Al Young met in the morning and devised a corrugated metal cooling liner that would fit inside the tailpipe. The cold airflow between the burner shell and liner kept the liner from burning up.[195] Silverstein recalled, "And you know, that cooling liner has been used ever since. That liner we probably knocked out in one day and stuck in there and ran it the next night. That was really a time of great creative development. We really moved ahead."[196]

The War Is Over

Following the last run of the GE I-40 on 3 August 1945, the AWT was shut down for two weeks of cleaning. During that period B-29 bombers dropped atomic bombs on two Japanese cities. The resulting destruction forced Japan to surrender and accept the terms of the Potsdam Declaration. The war was

Image 72: The Douglas XTB2D-1 Skypirate, an experimental torpedo bomber powered by the Pratt & Whitney R-4360 Wasp Major engine. The XTB2D-1 was too large to be used on pre-Midway carriers, and multiseat torpedo bombers were falling from favor by the time that the Midway carriers were put into action in late 1945. Carburetor air scoops and the effect of the R-4360's long propellers were investigated in the AWT in November and December 1944.[197] The propeller caused low-pressure recoveries in the engine inlet, particularly at high angles of attack.[198] (NASA C-1944-08060)

Image 73: A mechanic installs instrumentation on the Lockheed XR-60's R-4360-18 engine inside the closed AWT test section during a test for the navy. The October 1945 studies found that engine temperature did not exceed 450°F at takeoff or at cruising altitudes of 5,000, 10,000, or 15,000 feet. In this photograph, the four-bladed Curtiss propeller has been removed. (NASA C-1945-13598)

over. Despite German technological advancements, it was the piston-driven Allied air power that had been the decisive element in both Europe and Japan. The German Me-262 fighters were in the air in 1944, but their effect was minimized by a lack of strategic metals needed for both the engine and airframe. By the time that engineers had created alternative designs, the Allies had inflicted heavy damage on German manufacturers.[199]

In spite of the NACA's desire to deal exclusively with existing reciprocating engines, only two of AWT's eight wartime tests involved piston engines, the Wright R-3350 and the experimental torpedo bomber—the Douglas XTB2D-1 Skypirate. The Skypirate was powered by the Pratt & Whitney R-4360 Wasp Major engine, the country's largest mass-produced engine.[200] Immediately after Victory over Japan Day (V-J Day), the Republic YP-47M Thunderbolt

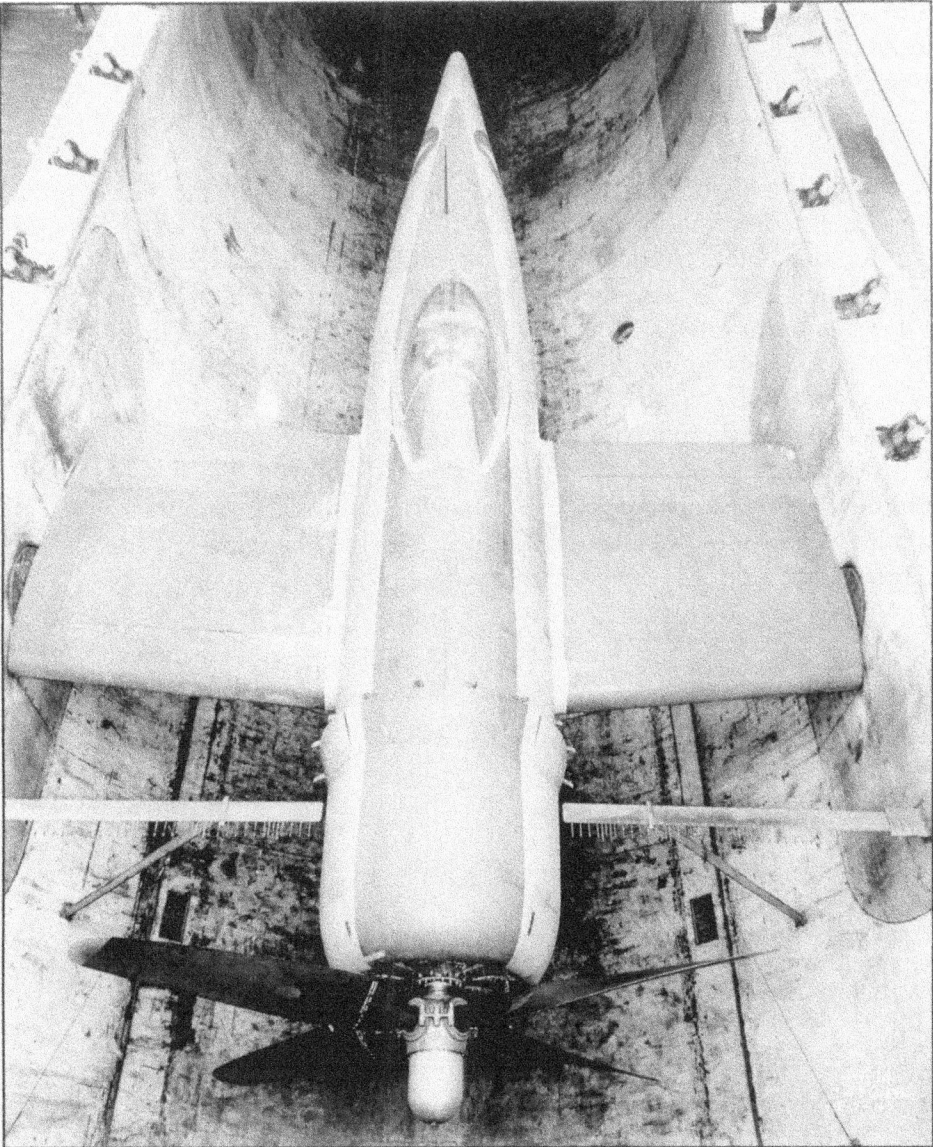

Image 74: After a two-week break in early August to clean up the tunnel and celebrate Victory over Japan Day (V-J Day), the AWT resumed testing on 17 August 1945 with the Republic YP-47M. The YP-47M, a faster variation of the P-47, was intended to combat the Messerschmitt jet fighters and defend against the V-2 bombs. Propellers from four different manufacturers were studied at altitudes up to 40,000 feet. The researchers developed curves to determine maximum efficiency and the distribution of the maximum thrust loading along the propeller blades during operation.[201] (NASA C-1945-13445).

and the R-4360 engine for Lockheed's XR-60 Constitution underwent brief analysis in the tunnel. Neither the Skypirate nor Constitution ever made it beyond the prototype stage. The YP-47Ms had seen some action in Europe in 1945 but continued to have problems when the war ended.

In the end, the AWT had little effect on the outcome of the war. Its most applicable contribution, the baffling for the B-29 engines, was not implemented until later. Although U.S. development of the turbojet would not be fast enough to make any significant impact on the outcome of the war, it did have long-term effects. The AWT demonstrated remarkable flexibility and became a significant contributor in developing the nation's first jet engines, both centrifugal and axial-flow. The AWT had not only survived the turbojet revolution, it had embraced it. During its first 18 months of service, the tunnel was modified to test jet engines, the staff showed that it could handle the new facility, a methodology of analysis was established that would be applied repeatedly over the next decade, and the emerging AERL leadership showed that it could change course with almost no warning.

Postwar Snapshot

After not taking a day of leave between the Pearl Harbor attack and the Armistice, the NACA Director of Aeronautical Research, George Lewis, would suffer two heart attacks during the first week of November 1945 and be forced into retirement two years later. From April to September 1945 Carlton Kemper participated in a team that followed in the wake of advancing Allied troops in Germany, collecting research papers, examining facilities, and interviewing researchers.[202] Kemper lost his position as Executive Engineer in the postwar reorganization, but he remained at the AERL as a consultant. Although Ray Sharp came to Cleveland to oversee the construction of the laboratory, his administrative skills were such that he would manage the lab until 1961. A former NACA Langley engineer, Jesse Hall, would serve as a technical liaison between Sharp and the research staff.

Abe Silverstein had been instrumental in testing the first jet engines and was already designing the lab's first supersonic wind tunnels. On 31 May 1945 Silverstein presented a summary of the recent engine studies conducted in the AWT at the GE Gas Turbine Conference. Despite having been aware of the turbojet for only a year and a half, the paper shows Silverstein's remarkable understanding of the gas turbine technology. He was able to summarize and integrate information from tests on GE's axial-flow TG-180 and centrifugal I-40 and on Westinghouse's 19B and 19XB engines.[203] Silverstein would spearhead the lab's research for the next 12 years.

Image 75: NACA representatives pose in front of the AWT. Below left to right: George Lewis, Carlton Kemper, and Ray Sharp. Above left to right: Abe Silverstein and Jesse Hall. (NASA C–1944–04955)

Endnotes for Chapter 4

167. Carlton Kemper, "Dr. Kemper's Talk at Morning Session of First Annual
     Inspection," Annual Inspection at NACA Lewis Flight Propulsion Laboratory,
     8–10 October 1947, NASA Glenn History Collection, Cleveland, OH.
168. Jeremy R. Kinney, "Starting From Scratch?: The American Aero Engine
     Industry, the Air Force, and the Jet, 1940–1960" (Reston, VA: American
     Institute of Aeronautics and Astronautics, 2003).
169. Richard Leyes II and William A. Fleming, *The History of North American
     Small Gas Turbine Aircraft Engines* (Reston, VA: American Institute of
     Aeronautics and Astronautics and Smithsonian Institution, 1999), p. 32.
170. Eugene Emme, *Aeronautics and Astronautics: An American Chronology of
     Science and Technology in the Exploration of Space, 1915–1960* (Washington,
     DC: NASA Headquarters, 1961), pp. 39–49.
171. William Fleming, *Altitude Wind Tunnel Investigation of Westinghouse
     19B-2, 19B-8, and 19XB-1 Jet Propulsion Engines, I: Operational Characteris-
     tics* (Washington, DC: NACA RM–e8J28, 1948), pp. 1–2.
172. Leyes, *The History of North American Small Gas Turbine Aircraft Engines,*
     pp. 36–39.
173. Fleming, "Altitude Wind Tunnel Investigation of Westinghouse Engines."
174. Leyes, *The History of North American Small Gas Turbine Aircraft Engines,*
     p. 39.
175. E.F.C. Somerscales and R.L. Hendrickson, "3500 kW Gas Turbine at the
     Schenectady Plant of the General Electric Company" (New York, NY:
     American Society of Mechanical Engineers, 1984), p. 6.
176. General Electric Company, *Seven Decades of Progress: A Heritage of Aircraft
     Turbine Technology* (Fallbrook, CA: Aero Publishers, Inc., 1979), p. 56.
177. Emme, *Aeronautics and Astronautics,* pp. 49–63.
178. Emme, *Aeronautics and Astronautics,* pp. 39–63.
179. Richard Golladay and Harry Bloomer, *Altitude Performance and Operational
     Characteristics of 29-Inch Diameter Tail-Pipe Burner With Several Fuel
     Systems and Fuel-Cooled Stage-Type Flame Holders on J35-A-5 Turbojet
     Engine* (Washington, DC: NACA RM–E50A19, 1950), p. 1.
180. Leyes, *The History of North American Small Gas Turbine Aircraft Engines,*
     p. 30.
181. Interview with Bruce Lundin, conducted by Tom Farmer, *This Way Up:
     Voices Climbing the Wind,* WVIZ Documentary, 1991, NASA Glenn History
     Collection, Cleveland, OH.
182. Kemper, "Dr. Kemper's Talk."
183. Abe Silverstein, "Altitude Wind Tunnel Investigations of Jet-Propulsion
     Engines," General Electric Gas Turbine Conference (31 May 1945), p. 13.
184. Silverstein, "Altitude Wind Tunnel Investigations of Jet-Propulsion Engines,"
     p. 10.
185. "The Afterburner Story," *Ryan Reporter* 15, no. 2 (1 April 1954): 2.

186. William A. Fleming, E.W. Conrad, and Alfred W. Young, *Experimental Investigation of Tailpipe Burner Design Variables* (Washington, DC: NACA RM–E50K22, 1950).

187. Fleming, "Experimental Investigation of Tailpipe Burner."

188. Interview with Abe Silverstein, conducted by John Sloop, 29 May 1974.

189. "The Afterburner Story," *Ryan Reporter.*

190. Bruce Lundin, David Gabriel, and William Fleming, *Summary of NACA Research on Afterburners for Turbojet Engines* (Washington, DC: NACA RM–E55L12, 1956).

191. Lundin, *Summary of NACA Research on Afterburners.*

192. "The Afterburner Story," *Ryan Reporter.*

193. Lundin interview, conducted by Tom Farmer, 1991.

194. Lundin interview, conducted by Tom Farmer, 1991.

195. Interview with Abe Silverstein, conducted by John Mauer, 10 March 1989, NASA Glenn History Collection, Cleveland, OH.

196. Silverstein interview, conducted by Sloop, 29 May 1974.

197. John Kuenzig and Herman Palter, *Aerodynamics of the Carburetor Air Scoop and the Engine Cowling on a Single-Engine Torpedo Bomber-Type Airplane* (Washington, DC: NACA MR E6E27, 1946).

198. NACA, "Thirty-Second Annual Report of the National Advisory Committee for Aeronautics" (Washington, DC: Government Printing Office, 1946), p. 27.

199. T.A. Heppenheimer, "The Jet Plane Is Born," *American Heritage: Invention and Technology Magazine* I, no. 2 (Fall 1993).

200. Kuenzig, Aerodynamics of the Carburetor Air Scoop.

201. Martin Saari and Lewis Wallner, *Altitude Wind Tunnel Investigation of Performance of Several Propellers on YP–47M Airplane at High Blade Loadings: V—Curtiss 836-14C2-18R1 Four Blade Propeller* (Washington, DC: NACA RM–E6J31, 1946), p. 9.

202. "Kemper Returns," *Wing Tips* (19 October 1945).

203. Silverstein, "Altitude Wind Tunnel Investigations of Jet-Propulsion Engines."

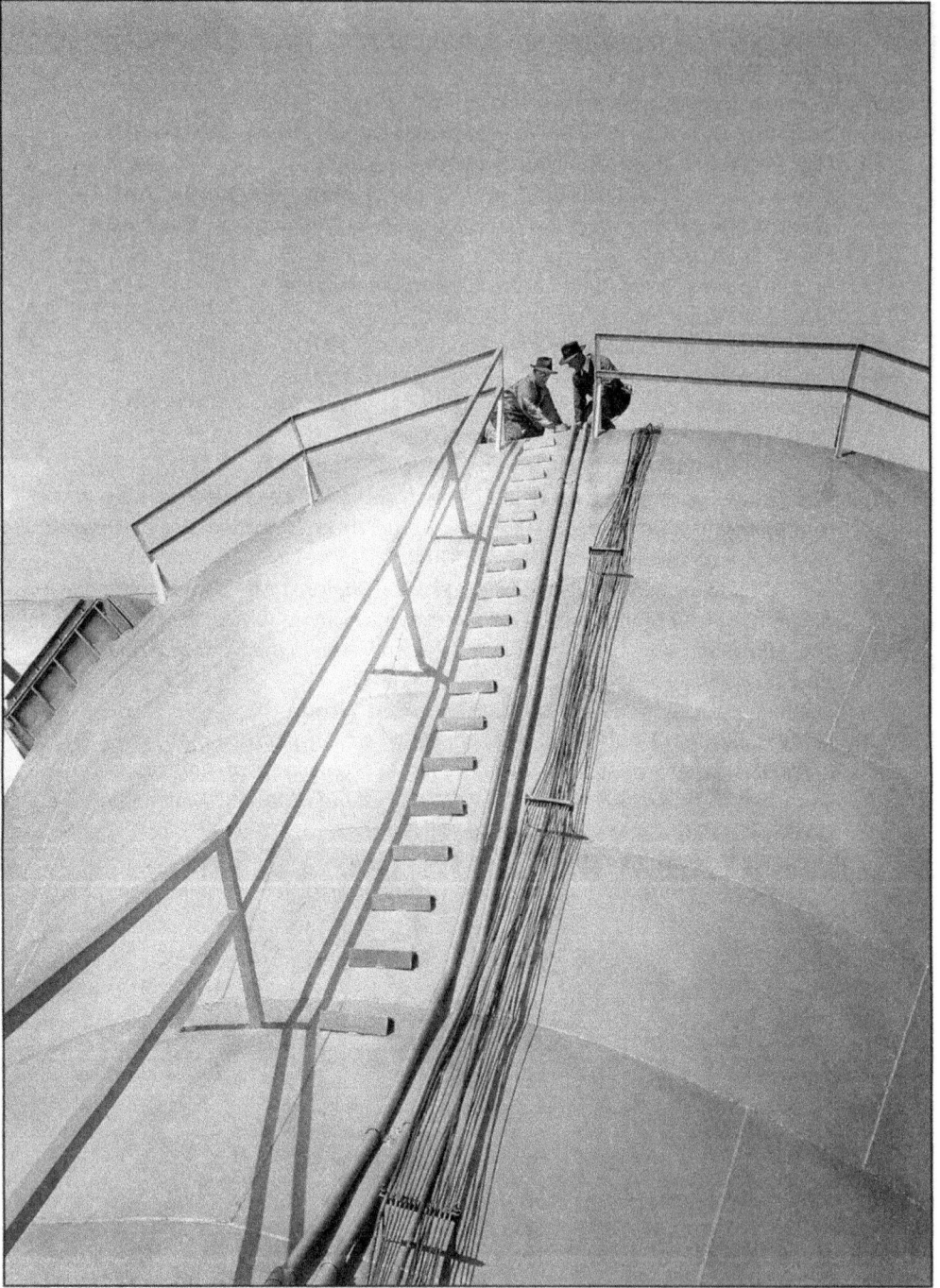

Image 76: Two men work on top of the 51-foot-diameter section of the Altitude Wind Tunnel. The facility emerged from the secrecy of the war and began a highly productive 10-year period of advancing turbojet technology. (NASA C–1945–10525)

Chapter 5

Will to Power | The Postwar Era (1946–1949)

"Trying to be a patriotic American I have refrained from taking this matter up with you during the war," wrote Robert Boone, owner of a nearby residence, "… almost every day except Sundays during the past year or more, the noise from your laboratory has been almost unbearable."[204] This 29 August 1945 complaint was one of the first indications that the respite following Victory over Japan Day (V-J Day) would be short lived. Although there were many sources of noise at the Aircraft Engine Research Laboratory (AERL), it was the inaudible low-frequency vibrations generated by the Altitude Wind Tunnel's (AWT) exhausters, which ran during the late night and early morning hours, that seemed to "have the greatest range of annoyance" for the surrounding community. By November the NACA had taken a number of steps to appease the neighbors, including the installation of mufflers over the vent pipes exiting the Exhauster Building.[205]

It was these AWT exhausters that were used to create the lab's two new supersonic wind tunnels. Now that the war had ended, the lab would have to grapple with the new field of high-speed flight, the return to basic research, and the reorganization of its staff. The rapid advancement in propulsion technology during the war years was unrivaled in the history of aviation. The AERL and AWT had tackled the changes on the fly and were poised to lead the way now that peace had settled. The AWT would continue the turbojet work it had started in 1944 while also studying ramjets, turboprops, and British engines. The benefit of the AWT's steady analysis of newly emerging jet engines such as the GE TG-190 and Westinghouse 24C would become evident with the powerful turbojets of the 1950s.

Image 77: The residence of Robert Boone, owner of the Store News Company, was built in the 1930s as a bucolic home in an undeveloped area along the Rocky River. Following the construction of the AERL just to the southeast in the early 1940s, Boone's home was besieged by noise from the test facilities, particularly the Propeller Research Test Facility and the AWT, which was less than a half mile away. The house was sold to the Guerin family in the 1950s but was acquired by NASA soon after as the lab expanded. NASA demolished the home in 2008. (1940s) (NASA C-1995-03926)

Image 78: During the final months of the war, the AERL began opening its doors to groups of writers, servicemen, aviation industry leaders, and others. Stands were built for group photos outside the lab's centerpiece, the AWT. On 5 July 1945, Fleet Admiral Erwin King headed a group of high-ranking officers who participated in an NACA and U.S. Navy research conference at the AERL. The group posed with NACA hosts George Lewis, Ray Sharp, John Victory, Jerome Hunsaker, and Addison Rothrock. (NASA C-1945-11156)

The Laboratory Takes a Moment To Regroup

The AERL staff was expanding quickly during this period. After doubling in 1944, the number of employees crested at 2,600 in early 1946. The staff would remain near that level until the early 1960s.[206] In October 1945, just after the end of the war, the staff was reorganized into four divisions to tackle its turbojet research efforts. Abe Silverstein was promoted to Chief of the new Wind Tunnels and Flight Division, and Al Young assumed the oversight of the AWT. The two would continue to share the same office in the AWT's Shop and Office Building for several years.[207] Silverstein's division formed a study group that met in the evenings and on weekends to discuss new technology and areas of research. For each session, one individual would study the literature on a specific area and teach the others.

Virginia Dawson's *Engines and Innovations* details the heated postwar debate regarding the influence of development on the lab. Some felt that the AERL was distancing itself from research, particularly with the use of full-size engines. The hurried nature of the wartime testing left little time to obtain basic data.[208] Silverstein, however, stressed that the NACA should strive to not provide just academic research but also to "keep it useful." Both the air force and the navy would remain active at the lab. Military representatives would meet frequently with the AERL staff to identify potential projects. The military would then submit a formal request to the NACA Headquarters for a test on a specific engine. Although the NACA acted as a service organization for the military and industry, Silverstein pushed AERL staff members to use their knowledge and experience to guide them into the best areas.[209]

The emergence of the jet engine required changes not only from the staff but also the test facilities. Not long after it came online, the AWT had been reconfigured to accommodate jet engines. Facility engineers found that by taking advantage of the pressure differences between the outside air environment and the pressures created by altitude simulation, they could provide enough pressure to simulate speeds that were one and two times the speed of sound. The ducting of the airflow directly to the inlet still allowed entire engines to be studied, but it effectively reduced the size of the test section so that fuselages and nacelle engine covers could not be included in the studies.[210] Although the AWT was adapted for the turbojet, Silverstein and other engineers were already designing other facilities for high-speed flight and larger engines.

Image 79: Construction of the Small Supersonic Tunnel behind the AWT in June 1945. This would be the lab's first supersonic tunnel. Eventually the building would house three small supersonic tunnels, referred to as the "stack tunnels" because of the vertical alignment. The two other tunnels were added to this structure in 1949 and 1951. (NASA C–1945–10764)

Image 80: Early drawing of the Small Supersonic Tunnels behind the AWT. The control room was in the basement. Half a story above was Tunnel No. 1, which had an 18- by 18-inch test section and could reach Mach 1.91. Tunnel No. 2 was a Mach 3.96 tunnel with a 24- by 24-inch test section. Tunnel No. 3, a Mach 3.05 tunnel, also had an 18- by 18-inch test section.[211] (NASA C–1946–14156)

Because the AWT only ran during the night, its exhausters sat idle most of the day. In the spring of 1945 Silverstein designed a small supersonic tunnel that utilized the AWT exhausters. The 2.25-square-foot-diameter open-circuit tunnel was built in just 90 days.[212] It was the first of three "stack" tunnels built just outside the AWT's southwest corner. Another small supersonic tunnel, referred to as the "Duct Lab," was created in the AWT's basement passage. These tunnels were small in size but yielded valuable data on high-speed aerodynamics. Bill Harrison, whom Silverstein pressured into running the stack tunnels, recalled, "We ran the living hell out of that thing. We really cranked them up."[213] Once these tunnels were working, Silverstein immediately took key members of the AWT design team aside to begin work on the large 8- by 6-Foot Supersonic Wind Tunnel. The 8×6, which was completed in 1948, would be the most significant of the NACA's postwar supersonic wind tunnels.

The AWT's unique ability to test full-size turbojets in altitude conditions resulted in an 8- to 12-month backlog of requests. To alleviate this problem, construction was begun on two altitude test cells in the Engine Research Building. This facility, referred to as the "Four Burner Area," contained static chambers into which full-size engines could be installed and run at altitudes up to 50,000 feet and temperatures ranging from 200 to −70°F. When it came online in 1947, it not only took some of the burden off of the AWT, but its compressors helped increase the capabilities of the AWT's exhaust system.[214] Work on a second, even larger, pair of engine test cells began almost immediately. When completed in 1952, this Propulsion Systems Laboratory would further reduce the AWT's workload.

Project Bumblebee

In 1944 the U.S. Navy's Office of Scientific Research and Development began developing a surface-to-air missile to combat Japanese Kamikaze attacks. The program, referred to as "Project Bumblebee," was based on the supersonic and long-range capabilities of the ramjet engine. The project was assigned to the Applied Astrophysics Laboratory at Johns Hopkins University. Although not used in the war, test flights continued throughout 1945.[215]

The ramjet system was extremely efficient at high speeds. Silverstein claimed in a 1951 interview, "The ramjet engine is more economical than the automobile engine in its use of fuel when flying at high supersonic speeds."[216] It was also the simplest type of propulsion engine. Like other engines, it was powered by combustion gases that were heated to high temperatures under pressure then exhausted. Compared with other engines, though, the ramjet was an extremely simple concept. It was basically a tube with no moving parts. Fixed grated devices, referred to as "flameholders," produced a constant flame to ignite the air passing through the ramjet.

The ramjet's major problem was that it could not operate until a certain airspeed was achieved, so either a turbojet or rocket had to be used to launch the vehicle. Another problem was fuel storage. The ramjet required a large quantity of fuel, but the aerodynamic design of supersonic missiles or aircraft generally do not provide much storage space.[217]

The idea of the ramjet had been around for years prior to Project Bumblebee, but component research and complete engine systems had yet to be completed. Industry did not have the massive quantities of process air needed to test full-scale ramjets, so the testing was taken on by the AERL. An NACA-designed 20-inch-diameter ramjet was installed in the AWT in May 1945. Thrust figures from these runs were compared with drag data from tests of scale models in small supersonic tunnels to verify the feasibility of

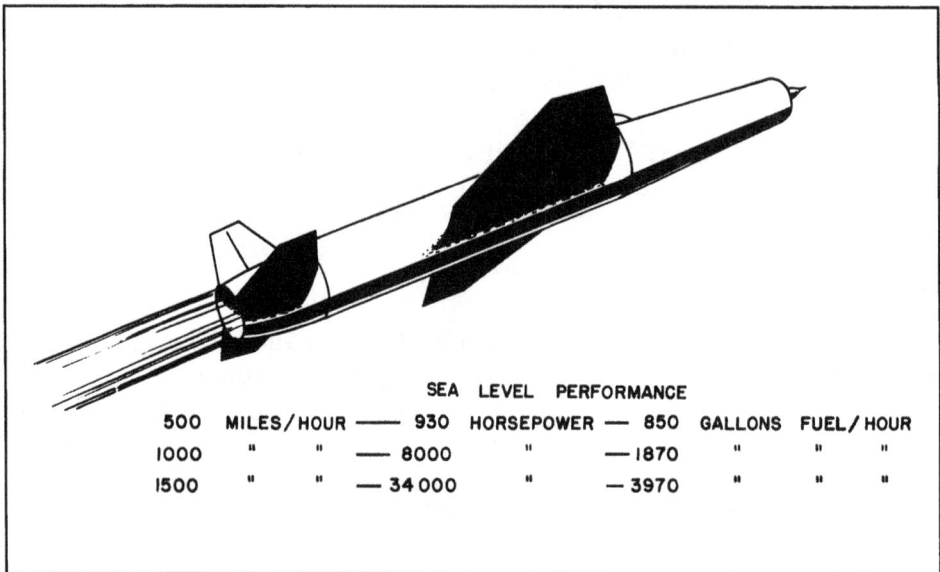

SEA LEVEL PERFORMANCE					
500	MILES/HOUR —— 930	HORSEPOWER — 850	GALLONS	FUEL/HOUR	
1000	" " — 8000	" — 1870	"	"	"
1500	" " — 34000	" — 3970	"	"	"

Image 81: The ramjet's potential for high-speed flight was unmatched by reciprocating, turbojet, or turboprop engines. The reciprocating engine handled air poorly, and the power of the turbojet and turboprop engines was limited by the temperature fatigue of the components. The advantage of the ramjet was its ability to process large volumes of combustion air, resulting in burning fuel at the optimal stoichiometric temperatures, which was not possible with turbojets. The higher the Mach number, the more efficient the ramjet operated.[218] (NASA C-1946-15566)

the ramjet.[219] The researchers found that an increase in altitude caused a reduction in the engine's horsepower. Optimal flameholder configurations were also determined for high-speed and high-altitude flights.[220]

The war ended before the Bumblebee missile was operational, but its development continued, and the scope of the project grew. By 1947 the ramjet diameter had been increased to 18 inches. The missile was no longer intended for just antiaircraft operations but also for long-range attacks on ground targets.[221]

In March 1947 Brigadier General Samuel Brentnall, Assistant Deputy Commanding General for Research and Development at Wright Field, requested that the NACA undertake a systematic study of the fundamentals of ramjets. He wrote, "The analytical problems of 'Why' and 'How', rather than 'What' is the important point."[222] The 8- by 6-Foot Supersonic Wind

Image 82: Mechanics working on the 20-inch NACA ramjet in the AWT test section. Referring to the ramjet tests in the AWT, Bill Harrison said, "Back in those days, about the only tool you needed was an inch and a quarter wrench to take the copper tubing apart. All this was, was a pipe with angle irons and fuel injectors."[223] The tunnel's refrigerated airflow was ducted directly into the ramjet's inlet (as seen in this photograph) to create the desired speed, static pressure, and temperature to simulate high-speed flight. The tunnel was used to analyze the ramjet's overall performance up to altitudes of 47,000 feet and speeds to Mach 1.84.[224] (NASA C-1946-14733)

Image 83: A 16-inch ramjet is installed in the AWT to study inlet shocks for the Project Bumblebee missile. The free jet air allowed the supersonic ramjet to be studied in the subsonic wind tunnel. The 18-foot-long ramjet was tested in the AWT from January through May 1949. Initially there was some concern with the number of contractors to be given clearance and allowed to witness the tests. A request from United Aircraft Corporation to have a representative present was declined.[225] (NASA C–1949–23409)

Tunnel would not be completed for another year, so researchers at the Cleveland lab resorted to flight-testing the ramjets. The ramjets were initially dropped from aircraft so that gravity would accelerate them to a speed at which the engines could operate.[226]

The combustion performance of the ramjet was difficult to study during the high-altitude drop tests, so an 18-inch-diameter version was tested in the AWT during the winter of 1947/48. The studies focused on variations of the flameholder and fuel mixtures. Because the tunnel's airflow was ducted directly to the inlet, the AWT was once again able to simulate supersonic speeds at an altitude of 30,000 feet. The optimal configuration was found to be a can-type flameholder with a kerosene-propylene oxide fuel mixture.[227] The free-flight investigations showed that a variable-area nozzle would be required for the ramjet to operate efficiently at different speeds and flight conditions. The AWT tests confirmed the increase in performance with the variable-area nozzle plug.[228]

The Bumblebee tests resumed in the AWT the following January under the guidance of Dr. Wilbur Goss. Dr. Goss had developed the combustion system for the missile at Johns Hopkins.[229] This time the tests were on a 16-inch-diameter, 18-foot-long ramjet. The tests were similar to those from the previous year—the air was ducted through a supersonic diffuser, and the focus was on flameholder and fuel configurations. The combustion efficiency for three flameholder designs was verified, and it was found that gasoline provided better efficiency than kerosene.[230]

Though its development was protracted, Project Bumblebee resulted in several important navy missiles. The first and largest was Talos, which became operational in 1958 just as the Cold War was heating up. The missile, with its 300-pound warhead, was used extensively in the Vietnam War before it was retired in 1979. Smaller Bumblebee missiles, such as the Tartar and Terrier, came into service shortly thereafter and continue to be used by the navy today.[231]

Image 84: "A Talos missile roars off the fantail of the Guided Missile Cruiser USS Galveston (CLG-3) in the Caribbean Sea. The deadly Talos, which has been in the testing stages for the past three years, was again proven operational last month when the Galveston, the ship that first fired the missile in 1958, made history again by completing the longest range Talos firing at sea." (Quoted from the original photo caption, released by the ship's Public Information Office on 17 March 1961.) (U.S. Navy NH 98846)

Photo Essay 2:
How Engines Were Tested in the Altitude Wind Tunnel

Photo Essay 2:
How Engines Were Tested in the Altitude Wind Tunnel

NACA Lewis Altitude Wind Tunnel

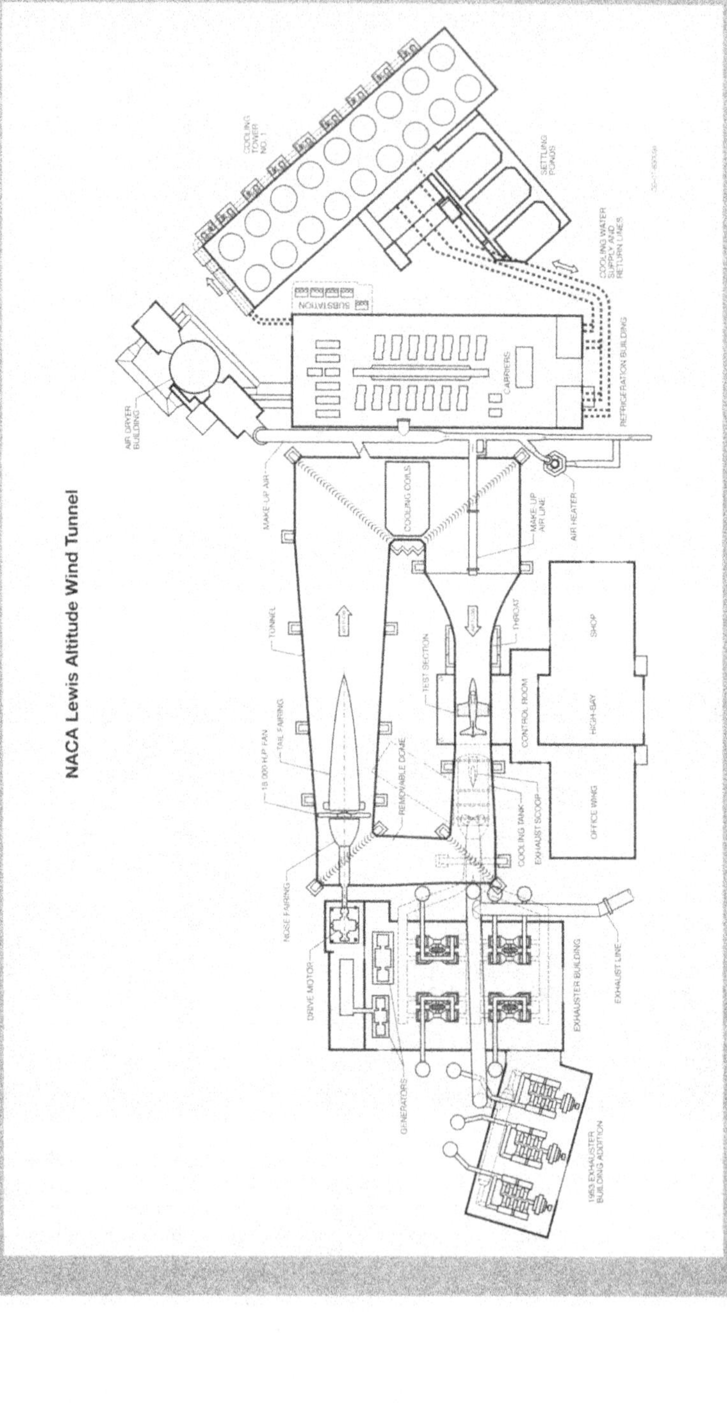

Image 85: Diagram of the AWT showing the many internal components of both the tunnel and its support buildings. (NASA CD-07-83009)

Image 86: The AWT complex, seen here with 1) the Icing Research Tunnel, included a number of support buildings: 2) the AWT, 3) the Refrigeration Building, 4) the Exhauster Building, 5) Cooling Tower No. 1, and 6) the Air Dryer Building. (NASA C-1944-05062)

Image 87: Research engineers developed ideas for tests that were often in response to requests from the military to improve a specific engine. Many of the researchers were located in the office wing of the Shop and Office Building. (NASA C–1956–43538)

Image 88: Arrangements were made to obtain an engine to test and to transport it to the Cleveland lab. The engine was brought into the AWT shop area, where it was readied for the tunnel. It was common for several different engines to be worked on simultaneously in the shop. (NASA C–1947–18019)

Image 89: *The researcher would discuss the engine and the test objectives with the Test Installation Division and the AWT technicians. The operations team would handle the installation of the instrumentation and fitting the test into the tunnel's schedule. Bill Harrison said, "Some of the engineers didn't particularly like to talk to us. They felt like 'You give me the mechanics and I'll get the work done.' I said, 'No, it doesn't work that way. You give me the work and I'll get it done.'"[252] Left to right: Bill Reiwaldt, Jack Wagner, and Dick Golladay. (NASA C–1956–43540)*

Image 90: *Upon completion of the previous test, the engine was removed. The next engine was lifted by an overhead crane and transported from the shop, up into the second story test chamber room, and finally into the 20-foot-diameter test section. The crane ran along the high bay from the shop to the test chamber. (NASA C–1951–28463)*

Image 91: The engine was mounted to a wing span that was fixed to trunions on the tunnel walls. The wing was connected to the measurement devices in the balance chamber. The tunnel mechanics connected the instrumentation to data output lines, installed the fuel and oil supply lines, and hooked up the engine's operating controls. (NASA C–1946–14244)

Image 92: Harold Friedman inspects the Toledo scales in the balance chamber. The balances were located in the balance chamber below the test section. Toledo scales recorded the movements of the balances caused by the engine being tested. The maximum thrust capacity was 12,000 pounds, and the maximum lift was 20,000 pounds.[233] (NASA C–1945–09635)

Image 93: The tunnel's clamshell lid was then lowered and fastened into place using hand-turned locking wheels. Windows along the side of the lid allowed photographers to film the test and observers to watch for engine failures. (NASA C–1951–28499)

Image 94: To start a test, a tunnel operator would activate the tunnel's drive motor, refrigeration system, and exhausters from inside the AWT control room. The operator worked in coordination with colleagues in the Engine Research, Exhauster, and Refrigeration buildings. (NASA C–1945–10346)

Image 95: The Carrier refrigeration equipment in the Refrigeration Building was activated to reduce the tunnel's air temperature to levels found at high altitudes. The building had its own control room and operators. It was the largest cooling system in the world when it was built and remains in use today. (NASA C–1944–06714)

Image 96: The massive exhausters in the AWT Exhauster Building interconnected with the Engine Research Building's exhaust system and also, after 1951, to the Propulsion Systems Laboratory's exhaust system. Originally the tunnel's atmosphere could be reduced to simulate a pressure altitude of 50,000 feet. This was later improved to 100,000 feet. (NASA C–1944–06710)

Image 97: The 31-foot-diameter, 12-bladed spruce fan in the southeast corner of the tunnel could generate wind speeds up to 500 mph depending on the pressure levels. The large tail fairing at the center of the fan helped to even the airflow. (NASA C–1944–03993)

Image 98: An 18,000-horsepower General Electric induction motor was used to rotate the AWT fan assembly. The motor was housed in the third story of the Exhauster Building. Its drive shaft exited the building then penetrated the tunnel wall and coupled into the fan assembly. Two large generators on the first floor were activated first to supply power to the drive motor. (NASA C–1947–18325)

Image 99: Turning vanes were located in each corner to straighten the airflow and direct it around the tunnel's square corners. The fan's drive shaft can be seen penetrating this set of vanes. (NASA C-1944-03991)

Image 100: The airflow passed through a bank of accordion-shaped cooling coils specially designed to reduce the AWT's temperature to −47°F. As the air passed through the banks of coils, heat was transferred to liquid Freon in the coils and carried away. (NASA C-1950-25465)

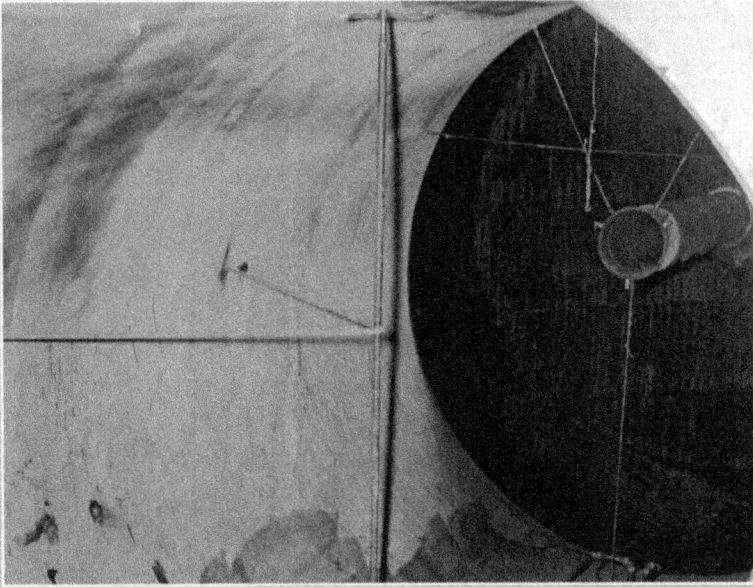

Image 101: This makeup air system pipe replenished the airflow with cool, dry air just upstream from the test section. This was atmospheric air that had been dehumidified and cooled in the Air Dryer Building outside the south-west corner of the tunnel. To increase the tunnel's capacity, the pipe was later ducted directly to the engine inlets. (NASA C–1951–27823)

Image 102: The tunnel contracted just prior to the test section to increase the velocity of the airflow. This "throat" section narrowed from 51 to 20 feet in diameter. (NASA C–1979–04019)

Image 103: *The engine was operated and tested inside the 20-foot-diameter test section. There were numerous viewing windows in both the lid and the bottom of the tunnel. (NASA C–1953–33230)*

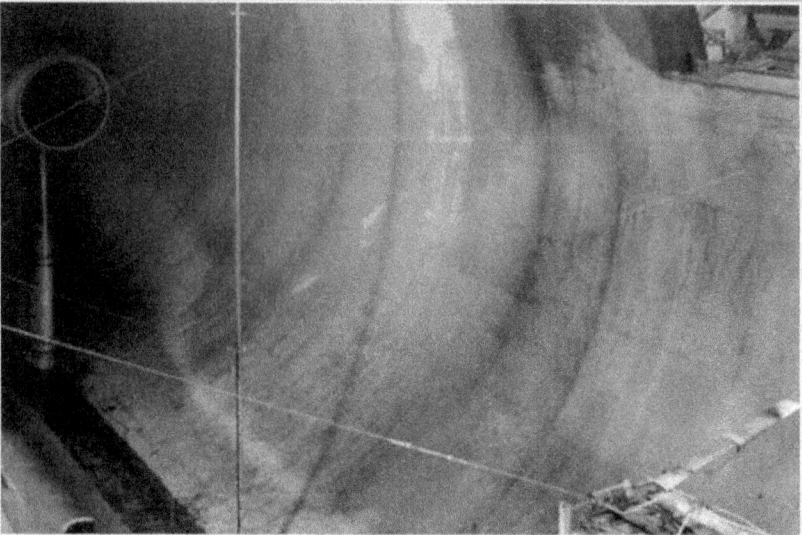

Image 104: *An air scoop, seen here at the left, was located just downstream from the test section to collect the combustion air exhausted by the engine. This 40-inch-diameter scoop stood statically in line with the engine's exhaust nozzle. The scoop was able to remove most of the engine's byproducts, thus preventing contamination of the airstream. (NASA C–1945–08709)*

Image 105: The researcher at this station in the control room controlled the engine in the test section. The soundproof control room was located in the balance chamber next to the lower portion of the test section. From here the researcher could run the engine at various power levels and monitor numerous aspects of the engine's behavior. (NASA C-1946-14241)

Image 106: Occasionally engines failed during tests, like the GE I-40 in this photograph. A failure could significantly damage not only the engine but also the wind tunnel. (NASA C-1945-11214)

16 W.T. TEST 2.5   RUN 3.1   MANO NO.2   M=.8   α=

Image 107: Female computers obtained test data from the manometers and other instruments and made the computations. Analysts then converted the data into meaningful figures. The "computers" and analysts were located in the Shop and Office Building office wing until being relocated to the 8- by 6-Foot Supersonic Wind Tunnel in 1948. Al Young's wife, Mercedes, served as the computers' supervisor for years. (NASA C-1944-04238)

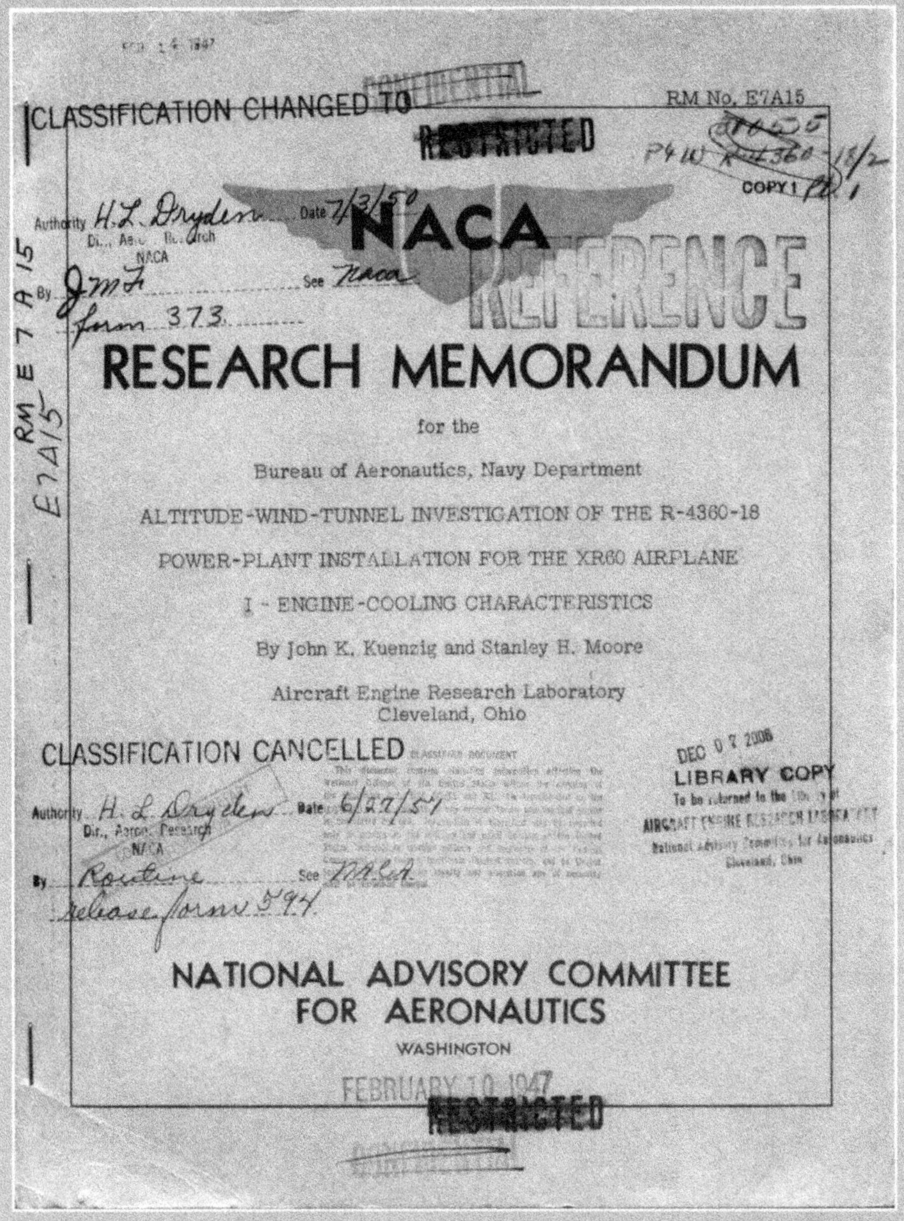
CLASSIFICATION CHANGED TO CONFIDENTIAL

RM No. E7A15

RESTRICTED

COPY 1

Authority H. L. Dryden Date 7/3/50

Dir., Aero. Research

NACA

By JMF See NACA

form 373

NACA

REFERENCE

RESEARCH MEMORANDUM

for the

Bureau of Aeronautics, Navy Department

ALTITUDE-WIND-TUNNEL INVESTIGATION OF THE R-4360-18

POWER-PLANT INSTALLATION FOR THE XR60 AIRPLANE

I - ENGINE-COOLING CHARACTERISTICS

By John K. Kuenzig and Stanley H. Moore

Aircraft Engine Research Laboratory
Cleveland, Ohio

CLASSIFICATION CANCELLED

Authority H. L. Dryden Date 6/29/54

Dir., Aero. Research

NACA

By Routine See NACA

release form 594

DEC 0 7 2006

LIBRARY COPY

NATIONAL ADVISORY COMMITTEE
FOR AERONAUTICS

WASHINGTON

FEBRUARY 10, 1947

RESTRICTED

CONFIDENTIAL

Image 108: Researchers analyzed the test results and compared them with the expected engine behavior or results from previous runs. Engines were tested over numerous runs under varying conditions and with variations on the configuration. The findings and test procedure were then described in research or technical memorandums and distributed to industry. (NACA RM No. E7A15)

"We Were All a Part of the Team"

Every test in the AWT required the coordinated effort of several different groups of people with complementary skill sets. The success of the AWT would have been impossible without this integrated team. The maintenance engineers or mechanics assembled and installed the test articles, operations engineers ran the tunnel, and researchers devised the tests and often operated the engines. "Computers" and analysts converted the raw data into usable form for the researchers. From the data and the physical observation of the previous tests, the researchers and technicians would make adjustments prior to the next set of runs. Abe Silverstein recalled, "To coordinate our next day activity, we'd have a meeting in the morning with the night crew, who worked from midnight on, would stay over, and we'd have a meeting with them when we came in. And then we'd decide what to do for the day."[234]

The research engineers played a vital role, but the contributions of the mechanics and technical staff were often unsung. During the NACA's most fruitful years, the mechanics and technicians not only operated the AWT and other large facilities but contributed to the research. Bill Harrison said in 2005, "You'll never see it happen again that way. Abe [Silverstein] felt we were part of the team and it took us all to design it, build it, install it, run it, and take the data."[235]

Mechanics could provide a different perspective on a problem. Howard Wine explained, "You had many, many instances where the engineers will come in with a proposal to run some tests that night and they didn't get the results they wanted [the previous night] for one reason or another. They say, 'well, we got to...change this through this nozzle area or do something. I really don't know exactly how to do this or what we ought to do in order to accomplish that.' And we [mechanics] would always say, 'well, why don't you go back to the office and come back at five and maybe we'll come up with something.' So they would go back to their office and we would start getting the hammers and cutting torches, welding machines and everything cranked up and make the modifications. They would come back and say, 'that's great, now wait a minute let me copy that down so I could put that on a drawing.' You know sometimes the drawing came after the fact."[236]

Harrison described the skill of a former technician, Bill Schwab, "This guy is the greatest inventor you've ever seen in your life." He recalled the Materials and Stress Division "had a bunch of physicists over there that knew what they wanted, but didn't know how to build it. They said, 'I want something that does this, this, and this.' This guy's a mechanic and [the physicist has] eighteen degrees, but he doesn't know how to do anything because they don't get taught any engineering courses. It's all physics, thinking, going to the Moon, how fast is the speed of light, and all that jazz. I don't care as long as the bulb turns on."[237]

There was some level of competition between the groups, but the one-on-one relationships were normally friendly, especially in the early days at the lab. The lab's management appreciated the fact that the groups needed one another to succeed and promoted comradery. Harrison recalled, "It [the early AERL] was a big family. We used to go to hockey games together. [Ray] Sharp and his wife, the whole gang. All the engineers, all the mechanics, everybody blended together."[238]

Image 109: A mechanic installs instrumentation on a 20-inch-diameter ramjet in the AWT test section. The technical staff was often able to troubleshoot testing problems and provide solutions for the research engineers. (NASA C-1946-14246)

Image 110: Researchers examine drawings and reports in an office in the Shop and Office Building. They worked closely with the technical staff to run tests in the AWT. (NASA C-1956-43539)

Westinghouse Hits Its Stride

Westinghouse and GE had made strides with their 19B and TG-180 jet engines, but it was their successors—the Westinghouse 24C (J34) and the GE TG-190 (J47)—that brought the success that the companies had hoped for when they jumped into the turbojet business. There was a good deal riding on these two new models, and between January 1947 and May 1951, they were studied for a combined 33 months in the AWT.

The 24C's 3,000-plus pounds of thrust doubled the power of the 19B. Variations were used on numerous aircraft from the late 1940s to the 1980s, including McDonnell's XF-85 Goblin, Douglas's D-558-II Skyrocket, and the NACA's X-3 Stiletto.

Although the navy had requested as early as February 1945 that the AERL study the 24C's performance characteristics, Westinghouse could not supply an engine until the war was over in the fall.[239, 240] After the reorganization in October, there was intense debate at the AERL about whether or not the lab should distance itself from commercial engines and concentrate on fundamental research. Kemper felt that the lab would also have to purchase additional expensive equipment to properly test the 24C. In March 1946 AERL Executive Engineer Carlton Kemper recommended that the pending Westinghouse 24C tests be cancelled.[241] When Wellington Hines, Director of the U.S. Navy's Bureau of Aeronautics, formerly requested the cancellation of AERL testing on 15 April, he stated that the previous studies of the 19XB engine in the AWT combined with Westinghouse's own tests were sufficient to forecast the 24C's performance.[242]

Image 111: William Fleming was one of the AWT's most significant researchers. He had been with the facility from the beginning and headed one of the tunnel's two research sections. He led investigations on every axial-flow jet engine studied in the AWT during the 1940s. (NASA C-1958-48622)

Less than a year later, however, a Westinghouse X24C-4B engine began an extensive nine-month examination in the AWT. It was among the most detailed investigations ever conducted in the facility. Overall engine performance, operating range, acceleration, deceleration, starting, and fuel consumption were studied. Specific problems such as windmilling, combustion blowouts, and afterburner configurations were also investigated at simulated flight Mach numbers and altitudes up to 50,000 feet.[243]

The test runs began in January 1947. By March, the engine had to be replaced because of two turbine failures. Westinghouse had designed the engine to operate with uneven temperatures and pressures so that the least-stressed portion of the compressor blades bore the highest temperatures. The AERL requested that Westinghouse provide these specifications and a second engine with a modified compressor and combustion chamber before

Image 112: A camera is set up in the AWT to film the combustion process in the afterburner of the Westinghouse 24C-WE-22 engine. Bob Walker recalled that, as a colleague was leaning onto the test section lid to monitor one of the Westinghouse runs through a portal, the engine seized—causing the turbine wheel to fly off, "We tended to doze up there on the midnight shift. He said, 'man, I thought I was going to jump right off the lid, that wheel came right past the window.' He was truly excited about that. It really wrecked that engine."[244] (NASA C-1951-27657)

proceeding with the tests. To avoid another failure, the AERL would not sub-ject the new X24C-4B to maximum engine speeds, altitude, or acceleration during ensuing runs.[245]

For the first time, icing tests were run in the AWT during the analysis of the X24C-4B. The primary concern was the overheating of the tailpipe burners because of ice buildup on the engine's inlet. William Fleming led an effort to draw some of the engine's high-temperature gas from the turbine inlet and to exhaust it just ahead of the compressor inlet. A stanchion with five spray nozzles was placed in the test section to create the icing conditions. Fleming and the mechanics could not get the spray system to create useful size droplets, and the tests were inconclusive. Even though the Icing Research Tunnel was experiencing the same difficulties with its droplets at that time, all future icing tests would be conducted there.[246]

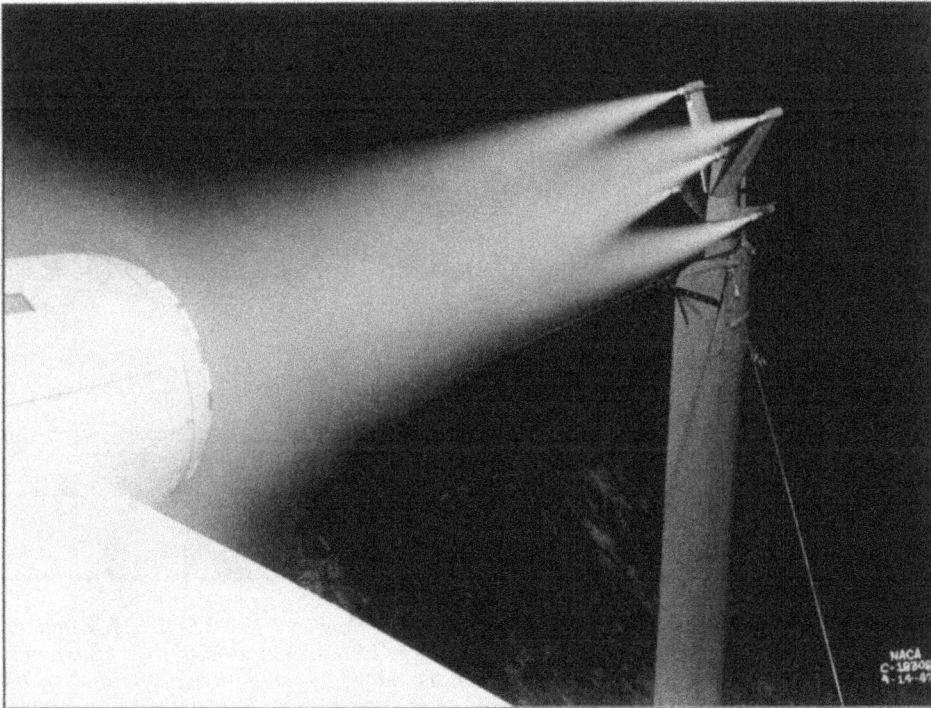

Image 113: A jet engine ice-protection system for the Westinghouse 24C-4B was studied at altitudes from 5,000 to 20,000 feet over a range of speeds in the AWT. Since icing tests were not usually conducted in the tunnel, this five-nozzle spray tower was installed just upstream from the inlet. The spray system turned out to be inadequate, so future hot-gas bleed tests were conducted in the Icing Research Tunnel. (NASA C-1947-18308)

In June 1950 the 24C (J34) returned for more than another year of test-ing in the AWT. The XJ34-WE-32 was the standard 11-stage, axial-flow 24C equipped with an afterburner and automatically controlled clamshell-type variable-area nozzle. The navy's Vought F7U Cutlass fighter jet, which was powered by two of these engines, had been suffering thrust problems.[247] Different variable-area nozzle configurations were tried, and a system using air to flush out the fuel-injection holes as the afterburner shut off seemed to improve the engine performance.[248-251] The AWT studies also addressed the new automatic thrust controller. This device consolidated many operational mechanisms used by the pilot into a single lever. It would also automati-cally control the afterburner. The interaction of the many parameters was extremely complicated. The AWT allowed the system to be checked out on a full-scale engine.[252]

Image 114: The Vought Cutlass was powered by two Westinghouse J34-WE-32 (24C) engines. In 1945 the navy had requested that a new fighter jet be designed around the 24C engines. The Cutlass was the first tailless military fighter, the navy's first swept-wing aircraft, and the first production aircraft to use afterburners. The first Cutlass is seen here at the NACA Langley Aeronautical Laboratory in December 1948. In just over a year, these engines would undergo a thorough investigation in the AWT. (NASA EL-2000-00267)

General Electric Takes Its Turn

GE had also reorganized after the war and decided to close its turbocharger and centrifugal compressor division in order to concentrate on their axial-flow engines. In March 1946 a team of young engineers began their first undertaking, a new 5,000-pound-thrust axial-flow engine. This TG-190 (J47) engine was to have the same physical size as the TG-180, but the new compressor, turbine, and lubrication system would produce an extra 1,000 pounds of thrust.[253]

The engine had not yet entered production when NACA Assistant Director of Aeronautical Research Russell Robinson requested that the TG-190 be installed in the AWT to study inlet-air efficiencies.[254] Between May 1948 and June 1950 the engine was tested three different times for a total of 15 months in the AWT. The studies were similar in nature to those being run on the Westinghouse 24C. The engine underwent an overall evaluation at various

Image 115: A technician measures one of the TG-190 engine's stator blades. The GE TG-190 (J47) was a successor to the TG-180. A different 12-stage axial-flow compressor and a single-stage turbine were installed in the TG-180 frame.[255] The TG-190 went through numerous studies in the AWT during 1948 and 1949. (NASA C–1949–22850)

Image 116: Left to right: John McAuley and William Prince install a 32-inch-diameter tailpipe on the GE TG-190 engine. McAuley and Prince determined that the airflow velocities were excessive at high speeds. They modified the engine's converging burner to reduce the burner inlet velocity and total pressure loss by 20 percent. They used a fixed conical exhaust nozzle, and the setup was at altitudes of 15,000 to 45,000 feet and speeds of Mach 0.22 to 0.76.[256] *(NASA C-1948-22168)*

speeds and altitudes up to 50,000 feet. Combustion chamber performance, high-altitude restarting, spark plug behavior, fuel distribution, and afterburner configurations were studied. The effect on variable inlet-air velocity and the use of alternative jet fuel also were tested.

The TG-190 was GE's most successful jet engine to date. Seventeen different models and over 30,000 engines were manufactured. They were used to power major military aircraft such as the Boeing B-47 and B-45, the North American F-86A, the Canadair Sabre, and the Convair B-36. In addition, the TG-190 would become the first axial-flow jet engine approved for commercial aircraft in the United States.[257]

The transition from reciprocating engines to turbojets took several different paths. One used a gas turbine to power a propeller. Several of these turboprop engines were studied in the AWT. The nation's first turboprop, the TG-100,

Image 117: GE's 14-stage axial-flow TG-100, the nation's first turboprop aircraft engine, underwent two months of testing in the AWT starting in late 1946. A 12-foot, 7-inch Hamilton Standard propeller was fitted to the engine. The TG-100 was run at 8,000 to 13,000 revolutions per minute (rpm) and at altitudes from 5,000 to 35,000 feet.[258] (NASA C–1946–17387)

had actually been developed before the turbojet. GE's steam turbine plant in Schenectady, New York, had been working on the turboprop for several years when it received the 1941 army contract to design an engine using an axial-flow compressor.[259]

The military was interested in an engine that would use less fuel than the early jets but would keep up performance-wise. The turboprop seemed to be the answer. Development of the engine was slow, however. It was eventually flown on a Convair XP-81 in December 1945, but engine problems persisted.[260]

A year later the TG-100 was tested for two months in the AWT for the air force. AERL researchers Lewis Wallner and Martin Saari put the engine through the usual AWT analysis and determined that its compressors, combustion chamber, and turbine were impervious to altitude variations. They were able to establish the optimal engine speed and propeller angle and to calculate performance figures at altitudes up to 35,000 feet based on operation of the engine at sea level.[261] Despite these findings, development of the TG-100 was cancelled in May 1947. Twenty-eight of the engines were produced, but they were never incorporated into production aircraft.[262]

British Engines Visit Cleveland

As the result of an informal March 1947 conference between representatives of the NACA and the British Ministry of Supply, NACA Lewis's Dr. Walter Olsen and several military and university representatives traveled to Great Britain in May. Almost exactly six years after the initial flight of the Whittle engine, this eight-person U.S. Gas Turbine Mission began a two-month survey of the British turbojet industry. Upon their return in July, the group visited the principal U.S. aircraft engine manufacturers, the U.S. Army's Wright Field facility, and the NACA lab in Cleveland. Olsen and the others briefed the staff at these facilities and incorporated their feedback into the final report comparing turbojet progress in Britain and the United States.[263]

Olsen and his group concluded that British engines were still more advanced than their U.S. counterparts, largely the result of Whittle's head start. The mission found that Britain's lack of large engine test facilities was starting to hamper their progress, though. Following Olsen's report, the NACA approached the U.S. Air Force for their assistance in obtaining British turbojet and turboprop engines to study in Lewis's facilities.[264]

The air force was especially interested in data comparing turboprop engine performance at sea level and altitude. The AWT and the two test cells in the Four

Image 118: Walter Olsen of Lewis's Fuels and Lubrication Division led a group of civilian, military, and NACA representatives on a 1947 fact-finding mission to investigate the progress of British turbojet manufacturers. Afterward several British engines were brought to Cleveland for testing in Lewis's altitude facilities. (NASA C-1956–41087)

Image 119: In response to a request from the NACA, Britain supplied an Armstrong-Siddeley contrarotating turboprop engine to study in the AWT. The tests ran from July 1949 through January 1950; it was the first time that the tunnel had been used to study an engine with the sole purpose of learning about, not improving, the engine. (NASA C–1949–23957)

Burner Area were the NACA's only facilities capable of conducting these studies.[265] A deal was reached that allowed several British engines to be examined at Lewis so long as it did not impede the testing of domestic engines and the resulting data could be shared with U.S. manufacturers.[266]

The Python

British automobile engine manufacturers began producing aircraft engines during World War I. This continued through the interwar period and into World War II. One of the most promising turbojets developed during the war was Rolls-Royce's Nene. After the war, the United States selected the Nene to be incorporated into its aircraft carrier fighters.[267] A Nene engine was brought to NACA Lewis in February 1948 for an intensive investigation in the Four Burner Area test cells.[‡]

The NACA had also requested several turboprop engines from the British. Originally they sought the Mamba, but by that time the engine was considered to be outdated. The next to be considered was the Proteus, but a delay in its availability led the NACA to select the Armstrong-Syddeley Python instead.[268] By January 1948, the U.S. Air Materiel Command informed the NACA that the British were phasing out their turboprop engines, so replacement parts for the Python might be in short supply during the NACA tests. The details were worked out, however, and the British began preparing a Python for shipment in April.[269]

The Python was a 14-stage axial-flow compressor turboprop with a fixed-area nozzle and contrarotating propellers.[270] Armstrong-Syddeley was another British automaker that expanded to aircraft engines during the interwar years. The Python was most successfully used on the Wyvern torpedo bomber, but the engine had been problematic. Although first flown in 1946, the Wyvern suffered numerous problems with its twin propellers. A redesigned version was flight tested in 1950, and the Wyvern eventually served the Royal Air Force from 1953 to 1959.[271]

Testing of the Python in the AWT finally began in July 1949. It was the first time that the tunnel was used for the sole purpose of learning about an engine, not trying to improve it. The Python's reverse-flow combustor coated the compressor blades with carbon and soot. If not removed, the soot would contaminate the tunnel's airflow. NACA retiree, Bob Walker, recalled the

---

[‡]Details of the Nene tests at Lewis are described in Dawson's *Engines and Innovation* (Ref. 208, Chapter 7).

Python, his first post-apprentice assignment. "I remember that first day when I went on the job. They gave us a bucket of barsol, which is a kind of kerosene derivative [and] some steel wool, really fine steel wool. [Another mechanic] and I were there cleaning this. We're cleaning all the blades and it's like a porcupine, you know. You had to be real careful that you wouldn't nick your self on it."[272]

Image 120: Left to right: Abe Silverstein and Martin Saari. Saari headed the AWT's second research section for most of the 1940s and 1950s. He was among the original staff taken on when Silverstein arrived in 1943. Saari's studies focused on the World War II piston engines, including the B-29's R-3350, and the TG-100 and Python turboprop engines. (NASA C-1958-47728)

The engine's dynamic response was studied using a frequency-response method at altitudes between 10,000 to 30,000 feet. Dynamic response is the way the engine performs while transiting from one steady unchanging condition to another. Determining these characteristics was a key element in perfecting engine designs. Seymour Himmel and Eugene Krebs led the team that discovered in the AWT that they could predict the dynamic response characteristics at any altitude from the data obtained from any other specific altitude. The dynamic response of the propeller and propeller/engine pairing were obtained during a single test run.[273]

End of the 1940s

As the 1940s drew to a close, the AWT completed its first sprint. The tunnel had been in almost constant operation for six years. It had played an important role in the steady development of the jet engine and was used to study ramjets and turboprops. Although the wartime investigations and the later Project Mercury and Centaur tests had higher profiles, the study and improvement of the axial-flow engine from 1945 to 1949 was the AWT's most enduring contribution to the aerospace field. The improvements developed in the 1940s would manifest themselves with drastic increases in engine performance and thrust in the early 1950s. Both the lab itself (which was renamed the "NACA Flight Propulsion Laboratory" in 1947 and the "NACA Lewis Flight Propulsion Laboratory" in September 1948) and the AWT would pause at this point to adjust themselves for the next era of propulsion.

Endnotes for Chapter 5

204. Robert L. Boone to Dr. George Lewis, 29 August 1945, NASA Glenn History Collection, Cleveland, OH.
205. Daniel Williams to Chief Engineer, Service Branch, 11 October 1945, NASA Glenn History Collection, Cleveland, OH.
206. "A Brief Picture of the Flight Propulsion Research Laboratory," 24 September 1947, NASA Glenn History Collection, Cleveland, OH.
207. "Research Reorganized," Wing Tips (19 October 1945).
208. Virginia P. Dawson, *Engines and Innovation: Lewis Laboratory and American Propulsion Technology* (Washington, DC: NASA SP–4306, 1991), chap. 2.
209. Interview with Abe Silverstein, conducted by Virginia Dawson, 5 October 1984, NASA Glenn History Collection, Cleveland, OH.
210. "History of a Wind Tunnel," Film C–216 (Cleveland, OH: NASA Lewis Research Center, 1961), NASA Glenn History Collection, Cleveland, OH.
211. "Major Research Facilities of the Lewis Flight Propulsion Laboratory, NACA Cleveland, Ohio, Wind Tunnels—Small Supersonic Wind Tunnels," 24 July 1956, NASA Glenn History Collection, Cleveland, OH.
212. "NACA Announces New Supersonic Wind Tunnel for Jet Propulsion Research," *Wing Tips* (11 August 1945).
213. Interview with Bill Harrison, conducted by Bob Arrighi, 14 October 2005, NASA Glenn History Collection, Cleveland, OH.
214. Carlton Kemper, "Dr. Kemper's Talk at Morning Session of First Annual Inspection," Annual Inspection at NACA Lewis Flight Propulsion Laboratory, 8–10 October 1947, NASA Glenn History Collection, Cleveland, OH.
215. National Air and Space Museum Collections Database, Space History Division, Smithsonian, "Rockets and Missiles: Bumblebee," Inventory Number: A19510009000, 18 August 1999, *http://collections.nasm.si.edu/code/emuseum.asp* (accessed 8 August 2007)
216. Interview with Abe Silverstein, conducted by J.J. Haggerty, 21 October 1951, for *American Aviation Magazine,* NASA Glenn History Collection, Cleveland, OH.
217. Abe Silverstein, "Ramjet Propulsion," c later 1940s, NASA Glenn History Collection, Cleveland, OH.
218. Silverstein, "Ramjet Propulsion."
219. NACA, "Thirty-Second Annual Report of the National Advisory Committee for Aeronautics" (Washington, DC: Government Printing Office, 1946), p. 26.
220. Silverstein, "Ramjet Propulsion."
221. National Air and Space Museum, "Rockets and Missiles: Bumblebee."
222. S.R. Brentall to George Lewis, 13 March 1947, NASA Glenn History Collection, Cleveland, OH.
223. Harrison interview, conducted by Bob Arrighi, 14 October 2005.
224. NACA, "Thirty-Second Annual Report."
225. J.W. Crowley to Edward Sharp, "Attendance of Contractor's Representatives During Tests of the Bumblebee Ramjet in the Altitude Wind Tunnel," 16 October 1945, NASA Glenn History Collection, Cleveland, OH.

226. Jesse Hall to Carlton Kemper, "Cleveland Laboratory Test Program for Investigating Ramjet Engines in Free Flight," 18 August 1947, NASA Glenn History Collection, Cleveland, OH.

227. W.H. Sterbentz and T.J. Nussdorfer, *Investigation of Performance of Bumblebee 18-Inch Ramjet With a Can-Type Flameholder* (Washington, DC: NACA RM–E8E21, 1948).

228. Robert McLarren, "NACA Lowers High-Speed Flight Hurdles," *Aviation Week* (9 May 1950).

229. Crowley to Sharp, "Attendance of Contractors."

230. T.J. Nussdorfer, D.C. Sederstrom, and E. Perchonok, *Investigation of Combustion in 16-Inch Ram Jet Under Simulated Conditions of High Altitude and High Mach Number* (Washington, DC: NACA RM–E50D04, 1948), p. 1.

231. Mark Wade, "Talos" (2001–2007), *http://www.astronautix.com/* (accessed 27 August 2009).

232. Interview with Bill Harrison, conducted by Bob Arrighi and Anne Burke, 9 August 2005, NASA Glenn History Collection, Cleveland, OH.

233. Arnold Biermann, "Technical Orientation Program for Engine Research Division Lecture No. VIII Facilities and Equipment," 1952, NASA Glenn History Collection, Cleveland, OH.

234. Interview with Abe Silverstein, conducted by John Sloop, 29 May 1974.

235. Harrison interview, conducted by Bob Arrighi, 14 October 2005.

236. Interview with Howard Wine, conducted by Bob Arrighi, 4 September 2005, NASA Glenn History Collection, Cleveland, OH.

237. Harrison interview, conducted by Bob Arrighi and Anne Burke, 9 August 2005.

238. Harrison interview, conducted by Bob Arrighi and Anne Burke, 9 August 2005.

239. George Lewis, "Investigation of the Performance Characteristics of the Turbine Component of the Westinghouse 24–C Jet Propulsion Unit" (research request), 17 February 1945.

240. George Lewis, "Request for Research Authorization to Test Westinghouse 24–C Turbine," 15 June 1945.

241. Carlton Kemper, "Proposed Cancellation of Westinghouse 24C Turbine Tests," 6 March 1946.

242. W.T. Hines, "Project TED–NACA–0421 24C Turbine—Cancellation of Request for Test On" 15 April 1946, "NASA Glenn History Collection, Cleveland, OH.

243. W. Kent Hawkins and Carl L. Meyer, *Altitude Wind Tunnel Investigation of Operational Characteristics of Westinghouse X24C–4B Axial-Flow Turbojet Engine* (Washington, DC: NACA RM–E8J25, 1948).

244. Interview with Bob Walker, conducted by Bob Arrighi, 2 August 2005, NASA Glenn History Collection, Cleveland, OH.

245. H. Sosnoski (Chief of Bureau of Aeronautics) to Bureau of Aeronautics Resident Representative, Essington, Pennsylvania, "Contract NOa(s) 3962 Turbojet Engine Calibration at AERL," 6 March 1947, NASA Glenn History Collection, Cleveland, OH.

246. William Fleming, "Hot-Gas Bleedback for Jet Engine Ice Protection," NACA Conference on Aircraft Ice Protection (1947), pp. 86–92.

247. Lewis Wallner and William Prince, *Investigation of Afterburner Performance and Afterburner Fuel System Coking of the Westinghouse* XJ34–WE–32 Engine (Washington, DC: NACA RM–E51H28, 1953).
248. Wallner, "Investigation of Afterburner Performance."
249. John McAuley, Adam Sobolewski, and Ivan Smith, Performance of the Components of the XJ34–We–32 Turbojet Engine Over a Range of Engine and Flight Conditions (Washington, DC: NACA RM–E51L10, 1952).
250. Harry Bloomer, William Walker, and George Pantages, Altitude Wind Tunnel Investigation of XJ3–WE–32 Engine Performance Without Electronic Control (Washington, DC: NACA RM–E51L12, 1952).
251. William Prince and J.T. Wintler, Investigation of Turbine-Outlet Temperature Distribution of XJ34–WE–32 Turbojet Engine (Washington, DC: NACA RM–E51K06, 1952).
252. David Novik and James Ketchum, "Controls Research," Triennial Inspection at NACA Lewis Flight Propulsion Laboratory, 9–11 October 1951, NASA Glenn History Collection, Cleveland, OH.
253. General Electric Company, *Seven Decades of Progress: A Heritage of Aircraft Turbine Technology* (Fallbrook, CA: Aero Publishers, Inc., 1979), pp. 59–60.
254. R.G. Robinson (Assistant Director of NACA Aeronautical Research) to Cleveland [AERL], "Effects of Velocity Distribution at Intake of Jet-Propulsion Engines," 21 May 1948, NASA Glenn History Collection, Cleveland, OH.
255. Carl Meyer, *Altitude Wind Tunnel Investigation of AN–F–58 Fuel in Experimental Version of J47 Turbojet Engine* (Washington, DC: NACA RM–E8L13, 1949).
256. William Prince and John McAuley, *Altitude Performance Characteristics of Tailpipe Burner With Converging Conical Burner Section on J47 Turbojet Engine* (Washington, DC: NACA RM–E50G13, 1950).
257. "General Electric J47 Turbojet Engine," National Museum of the United States Air Force, *http://www.nationalmuseum.af.mil/factsheets/factsheet.asp?id=6472* (accessed 27 August 2009).
258. Robert Geisenheyner and Joseph Berdysz, *Preliminary Results of an Altitude Wind Tunnel Investigation of an Axial Flow Gas Turbine Propeller Engine, III—Pressure and Temperature Distributions* (Washington, DC: NACA RM–E8F10b, 1948), p. 2.
259. E.F.C. Somerscales and R.L. Hendrickson, "3500 kW Gas Turbine at the Schenectady Plant of the General Electric Company" (New York, NY: American Society of Mechanical Engineers, 1984), p. 6.
260. National Museum of the United States Air Force, "General Electric T–31 Turboprop," *http://www.nationalmuseum.af.mil/factsheets/factsheet.asp?id=877* (accessed 27 August 2009).
261. Lewis Wallner and Martin Saari, *Altitude Investigation of Performance of Turbine-Propeller Engine and Its Components* (Washington, DC: NACA RM–E50H30, 1950), p. 14.
262. "General Electric T–31 Turboprop."

263. Walter T. Olsen to Carlton Kemper, "Tour of Aircraft Gas Turbine Industry in England by Walter Olsen," 22 August 1947, NASA Glenn History Collection, Cleveland, OH.

264. J.W. Crowley (NACA Associate Director of Aeronautical Research) to E.M. Powers (USAF Major General), 9 September 1947, NASA Glenn History Collection, Cleveland, OH.

265. E.M. Powers to George Lewis, 5 November 1947, NASA Glenn History Collection, Cleveland, OH.

266. Hugh Dryden to Chief of Bureau of Aeronautics, "NACA Investigation of British Gas Turbine Power Plants," 2 January 1948, NASA Glenn History Collection, Cleveland, OH.

267. T.A. Heppenheimer, "Rolls Royce and Its Aircraft Engines," US Centennial of Flight Commission, *http://www.centennialofflight.gov/essay/Aerospace/Rolls-Royce/Aero54.htm* (accessed 27 August 2009).

268. Edward Sharp to Hugh Dryden, "NACA Investigation of British Gas Turbine Type Power Plants." 24 November 1947, NASA Glenn History Collection, Directors Collection, Cleveland, OH.

269. R.J. Minty (USAF Colonel and Chief of Power Plant Lab) to Hugh Dryden, "NACA Investigation of British Gas Turbine Power Plants," 21 January 1948, NASA Glenn History Collection, Cleveland, OH.

270. Emmert Jansen and John McAulay, *Compressor Performance Characteristics of a Python Turbine-Propeller Engine Investigated in the Altitude Wind Tunnel* (Washington, DC: NACA RM–E50K24, 1951).

271. military-aircraft.org.uk, "Westland Wyvern Fighter Plane" (1 May 2005), *http://www.military-aircraft.org.uk/other-fighter-planes/westland-wyvern-fighter.htm* (accessed 29 December 2009)

272. Interview with Bob Walker, conducted by Bob Arrighi, 2 August 2005.

273. Eugene Krebs, et al., *Dynamic Investigation of Turbine-Propeller Engine Under Altitude Conditions* (Washington, DC: NACA RM–E50J24, 1953).

Image 121: Westinghouse's 25-foot-long, 11-stage axial-flow-compressor J40 turbojet underwent an 11-month period of study in the AWT. The AWT was upgraded in 1952 to handle these newer, more powerful jet engines. (NASA C–1951–28738)

Chapter 6

A Period of Transition

Big Engines to Small Rockets (1950–1958)

Construction of the new 8- by 6-Foot Supersonic Wind Tunnel was wrapping up when in September 1948 news arrived that the Soviet Union had installed a Communist government in North Korea. In an attempt to gain control over the entire peninsula, the new communist regime began supporting an insurgency in South Korea. When the insurgency appeared to be failing, the North Korean army undertook a predawn attack into the south on Sunday, 25 June 1950. A United Nations resolution was passed, and a U.S. naval fleet responded the following day with force. Several days later, the United States began moving ground troops into the fray.[274] The Korean War was the first armed conflict of the Cold War and the first opportunity for the jet aircraft to prove itself.

In the midst of these events, the NACA Lewis Flight Propulsion Laboratory reorganized again in September 1949. Abe Silverstein, who had moved his office from the Altitude Wind Tunnel (AWT) to the 8- by 6-Foot Supersonic Wind Tunnel (8×6) the year before, became Chief of Research for the entire lab. Upon assuming the vacant position of Associate Director in 1953, he finally moved to the Administration Building. As Associate Director, Silverstein continued oversight of the lab's research but also managed all operations, facilities, and construction. He claimed that his responsibilities and freedom had not changed much, "It just provided a notch in the organization chart that was filled by someone."[275]

The AWT Branch became part of the Engine Research Division.[276] Al Young was the AWT Branch Chief; David Gabriel, who would later head the Centaur Program, was Assistant Chief; and Carl Meyer, Martin Saari, and Reece Hensley led the three sections.[277] Although new facilities such as the 8×6, the Propulsion Systems Laboratory, and later the 10- by 10-Foot Supersonic Wind Tunnel would take on some of the AWT's propulsion work, the AWT was upgraded so that it could test the newer and increasingly larger turbojets. The nature of the AWT tests would change dramatically in the late 1950s, but for several years, the AWT continued the trends of the 1940s, such as a new Westinghouse turbojet, a turboprop, and a British jet engine.

Image 122: The swept-wing F-86A Sabre became the preeminent fighter of the Korean War. It was powered by the General Electric J47-GE-13 (TG-190), which had undergone intensive studies in the AWT during the late 1940s. The Sabre could reach an altitude of 49,000 feet and speeds of 685 mph, and it had a 10-to-1 kill advantage over the Russian MiG-15.[278] (1955) (NASA E-3996)

Image 123: Apprentice Dominic Giomini examines a Westinghouse 24C-7 that failed during a run in the AWT on 19 July 1950. That same day NACA Director Hugh Dryden had issued a letter proposing the acceleration of research at the lab in response to the outbreak of the Korean War three weeks before. (NASA C–1950–26297)

Upgrading for Big Engines

Some improvements were made to the turbojet engine in the postwar years, but its capabilities advanced exponentially from 1948 through 1955.[279] The AWT had made the transition from large reciprocating engines to the original jet engines rather easily, but the AWT's infrastructure was stressed by the size of the new turbojets. It was not a great leap from Wright's R-3350 3,500-pound-thrust piston engine to General Electric's (GE's 4,000-pound I-40 turbojet. New engines, such as Pratt & Whitney's J57, however, produced 13,500 pounds of thrust.

With the flare-up of hostilities in Korea, the NACA undertook a number of measures to keep up with new aeronautical advances. One was to upgrade the AWT. During the second half of 1951, a number of modifications were made to modernize the facility. The most dramatic change was the addition to the Exhauster Building. The small rectangular annex attached to the Exhauster Building's northeast corner housed three new Ingersoll-Rand compressors. A pump house and exhaust cooler pit were constructed underneath the tunnel, and two new cells were added to the cooling tower.[280] The modified AWT continued with its mission to analyze jet engines in the 1950s, although the engines were much larger than those studied several years before.

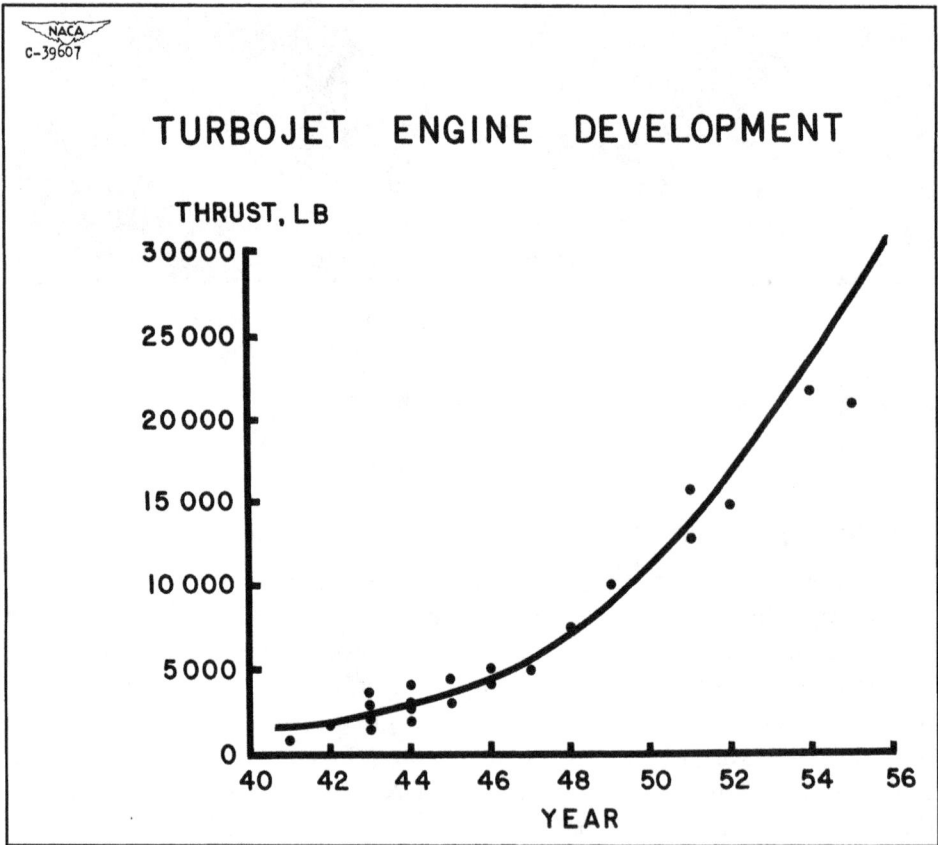

Image 124: Growth of the jet engine's capacity rose exponentially in the early 1950s. The first GE centrifugal engines produced 1,600 pounds of thrust in the early 1940s. (NASA C-1955-39607)

Photo Essay 3:
The Altitude Wind Tunnel Gets a Boost

Image 125. An addition was built on the AWT's Exhauster Building in 1951. This new wing contained three Ingersoll-Rand eight-cylinder reciprocating pumps. The original exhausters were initially constructed to handle 7 pounds of air per second at an altitude of 50,000 feet and 51 pounds per second at 28,000 feet. The new exhausters were upgraded to 12 pounds per second at 50,000 feet and 66 pounds per second at 28,000 feet. (NASA C–1953–31751)

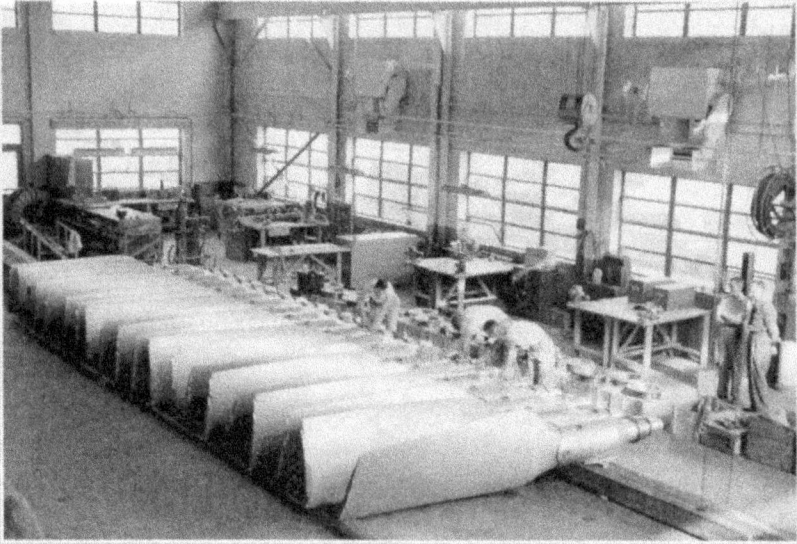

Image 126: The AWT's fan blades occasionally became worn or broken and had to be replaced. A new set of blades was installed on the 31-foot-diameter drive fan in 1951. The blades were prepared in the shop area, seen in this photograph, before being lowered into the AWT through a hole in the top of the tunnel and attached to the drive shaft. (NASA C-1951-28240)

Image 127: Workers install a large hub as part of a large overall upgrade of the AWT. The blades, hub, and fairings for the drive fan were replaced. The new fairings were roughly twice the original size. (NASA C-1951-28286)

Image 128: The AWT modernization included the installation of a new exhaust gas cooler manufactured by Biggs Company, in Akron, Ohio. This photograph shows the large cooler being moved between the Shop and Office Building to the left and the Exhauster Building to the right. The cooler was installed below the tunnel's air scoop. A new 60-inch-diameter scoop was also installed. (NASA C–1951–28229)

Image 129: A pump house was built underneath the northeast leg of the tunnel. This Circulating Water Pump House drew cooling water to be used by the new exhaust system. The pump house contained four Ingersoll Rand pumps, two 250-horsepower discharge pumps, and two 300-horsepower spray pumps. Another 75-horsepower spray pump was located in the northeast corner. These pumps drew and returned water from the Cooling Tower. (1951) (NASA C–2007–02591)

Westinghouse Falters

The J40 was one of the biggest disappointments in Westinghouse's thriving history. The 11,000-pound-thrust engine was intended to double the power of the successful 24C (J34). More specifically, it was intended to power the navy's new post-World-War-II fighter jets, such as the McDonnell F-3H Demon.

Westinghouse successfully operated the J40 in November 1948, but it was not until January 1951 that the J40 engine completed its 150-hour navy qualification test. The afterburner version would not qualify for another six months.[281] In the year following its initial successful flight on 7 August 1951, the F-3H Demon suffered eight major crashes; three were directly linked to the J40 engines.[282] From September 1951 through mid-July 1952, during the midst of these delays and quality concerns, a series of J40 investigations were conducted in the AWT.

Image 130: NACA Lewis mechanics prepare the 11-stage compressor of the Westinghouse J40 engine in the AWT shop area. (NASA C-1951-28841)

The navy sought quick modifications that could be easily incorporated into the J40 engines already in the field, as well as long-term solutions to prevent problems with engines in production. The XJ40-WE-6 was found to have serious compressor surge at high speeds. McDonnell had experienced this problem for the first time in a prototype F3H just as AWT researchers discovered it. Experiments in the AWT showed that changes to the angles of the rotor blades and new stator diaphragms solved the problem but reduced the thrust unacceptably. Use of a mixer at the compressor outlet, however, did improve the surge limit.[283]

The compressor and combustor were revamped, and the engine was redesignated the "J40-WE-8." It then underwent the general AWT checkout focusing on the effect of altitude on restarting, windmilling, and tailpipe burners.[284] The use of an adjustable turbine nozzle, as well as an adjustable

Image 131: A technician works on the 9,700-pound-thrust Allison J71-A-11 engine in the AWT test section. The 16-stage axial-flow engine was 16 feet long, 4 feet in diameter, and weighed 4,450 pounds. The J71 replaced Westinghouse's J40 on several navy fighters. (NASA C-1955-39623)

Image 132: The Rolls-Royce Avon RA14 engine was a 16-stage axial-flow compressor engine capable of producing 9,500 pounds of thrust. The Avon replaced the successful Nene engine in 1950 and remained in service until 1974. Lewis studied the RA-14 turbojet engine for the navy in the Four Burner Area in 1955 and in the AWT for 11 months in 1956. The engine was mounted on a stand capable of gauging engine thrust, and the tunnel's air was ducted to the engine through a venturi and bellmouth inlet, seen in this photograph.[285] (NASA C-1956-42786)

exhaust nozzle, increased both fuel efficiency and thrust.[286] Westinghouse continued to have problems fabricating the engines, and McDonnell recommended replacing them with the Allison J71, which two years later would also be tested in the AWT. In 1955 Westinghouse cancelled the J40, and the company, which 12 years before had built the first U.S. axial-flow turbojet, left the aircraft engine business altogether. The navy did switch to the J71, and the F-3H became one of the navy's primary fighters until the early 1960s.[287]

Image 133: A 16-stage Allison J71 jet engine is prepared in the shop for test runs in the Altitude Wind Tunnel. The J71, which was 16-feet long and 4-feet in diameter, could generate 9,700 pounds of thrust. (NASA C–1952–30101)

Image 134: The Allison T-38 turboprop was tested in the AWT from March to November 1953. (NASA C–1953–33216)

Apprentice School

To facilitate the close interaction of the lab's engineers, mechanics, technicians, and scientists, Ray Sharp established a four-year apprentice program to train craftsmen on a particular trade and basic scientific principles. The apprentice school covered a variety of trades, from aircraft mechanic to electronic instrumentation, machinist, and altitude systems mechanic. Abe Silverstein described the program as "the backbone of our work force, the core of skill which translates the engineer's and scientist's ideas into practical experiments and proven tests."[288]

The school was established in 1942, but faltered when over 90 percent of its students entered into the military. After the war, 40 of the original members returned to the lab. In some cases they were bumped to journeymen positions because of training received in the military. Bill Harrison was one of those who left for the military. He remembered returning to the program, "And they graduated us—though we had been broken up. We had been in the army

Image 135: An apprentice in the altitude systems program is shown learning the system in the Four Burner Area. The specialized skills required at NACA Lewis meant that apprentices were held to a higher standard than those in industry. They first had to pass written civil service exams. Previous experience with mechanical model airplanes, radio transmission, six months of work experience, or one year of trade school was required. (NASA C–1954–34953)

and navy—all over the world. We came back, and we had training in the army or navy for something. Of course, I learned how to drop bombs—that helped me a lot. But they decided to graduate our class just to show the next class that this is what you have to look forward to—graduation."[289]

That honorary first class in 1949 had only 15 graduates, but the number steadily increased to 45 with the next class in 1952 and to 110 in 1957.[290] There were over 600 graduates by 1969.[291] The program remained strong for decades. Many of the laboratory's future managers began their careers as apprentices including Bill Harrison, Bill Schwab, Andy Szuhai, George Tunder, and Bob Walker. Melvin "Lefty" Harrison, father of Bill Harrison and one of the original NACA Langley transfers, managed the program for several years in the early 1950s. The curriculum was created by the lab's skilled tradesmen. Silverstein told the 1952 class that it should strive to become involved in the larger research team, constantly improve its skills, and work to become leaders.[292]

Image 136: Members of the Apprentice Program's 1955 graduating class. The entire school and their families turned out for the ceremonies. Ray Sharp, Abe Silverstein, and Charles Herrmann were regular participants. Visitors such as John Victory, Hugh Dryden, Addison Rothrock, Henry Reid, or local officials often made speeches. (NASA C–1955–38393)

Walker recalled that an older high school classmate had been admitted into the Apprentice Program the previous year. "He thought he'd died and gone to heaven, because now he was getting paid to make model airplanes and other models...fancy equipment and all that. So he was telling me a little bit about the center, and I was interested in doing mechanical stuff. I asked him if they had any jet engines out there, and he said, 'Oh yeah, they've got jet engines.' And I said, 'Well, can you see them and everything?' And he said, 'Oh, you can take them apart, do anything you want to them. They'll help you take them apart.' I thought, boy, I could do that. I wouldn't need money to do that."[293]

The program, which was certified by both the Department of Labor and the State of Ohio, included classroom lectures, study of models, and hands-on work. The apprentices rotated through the various shops and facilities to provide them with a well-rounded understanding of the work at the lab. One hundred fifty of the 2,000 hours of annual training were spent in the classroom. A series of examinations was coupled with evaluation by supervisors in the shops. The apprentices were promoted through a series of grades until they reached journeyman status. Those who excelled in the Apprentice Program would be considered for a separate five-year engineering draftsman program. The 10 to 20 percent in each class that lagged behind were either reassigned or released from the NACA.[294]

In the mid-1960s a special effort was undertaken by the lab to reach out to inner-city high school students. A Pre-Apprentice Program was established in which potential candidates from the schools went through an intensive three-week training session. The top performers were given slots in the Apprentice Program at Lewis, while others used the experience to gain industrial apprenticeships. The program was a big success and continued into the 1990s.[295] Also in the 1960s, a separate apprentice program was established at Plum Brook Station.

The Mighty J57

The most powerful and successful of the postwar jet engines was the Pratt & Whitney J57. It was the first engine to exceed 10,000 pounds of thrust and was used to power the military's premier aircraft in the 1950s and 1960s, including the B-52 Stratofortress, the F-100 Super Sabre, the Lockheed U-2, and the F-102. A variant of the J57, the JT3, was used on Boeing's 707 and Douglas's DC-8 commercial airliners.

Unlike the other piston engine giant, Wright Aeronautical, Pratt & Whitney sought to carve out a space in the postwar turbojet market. At the request of the navy, the company bought Rolls-Royce's successful Nene jet engine in 1948. Pratt & Whitney used the Nene to create the 5,000-pound-thrust J42 and the 8,750-pound-thrust J48 in 1948 and 1949.[296] The breakthrough came two years later when the 13,500-pound-thrust J57 was first flown.[297]

The J57 was unique in that it had a dual axial-flow compressor that allowed it to break through the barrier that seemed to exist at 10,000 pounds of thrust. The engine employed two coaxial compressors, corresponding coaxial turbines, and a fixed-area nozzle. The inner spool had a seven-stage

Image 137: NASA's NB-52A with an X-15 underneath its wing. The B-52 was designed by Boeing in the late 1940s as a successor to the B-47 bomber and remains an important U.S. strategic bomber. Despite the design similarities, the B-52, powered by four pairs of Pratt & Whitney J57 engines, was a much heavier and faster aircraft. In early 1951, the Air Materiel Command requested the production of 13 B-52As, but in June 1952, 10 of the orders were changed to an updated version, the B-52B. One of the three B-52As, renamed the "NB-52A," became the mothership for the NACA's X-15 program. (1960) (NASA E-88-0180-4)

Image 138: The Pratt & Whitney J57 was tested extensively in the AWT from December 1953 to February 1954. General performance characteristics, inlet pressure distortions, and combustion efficiency were studied. Near the end of the investigations, one of the J57-P-1 engines suffered a major failure, seen in this photograph. (NASA C–1954–37208)

axial-flow compressor and a single-stage turbine. The outer spool had a nine-stage axial-flow compressor and two-stage shrouded turbine.[298]

The original version of the engine, the J57-P-1, was studied in the AWT in late 1953 during the run-up to the first flight of the B-52 bomber. The B-52 had been significantly redesigned in 1949 to include the then-secret J57 engines. AWT researchers studied the engine's performance and focused on flow ejectors, bleed ports, fuel flow, and inlet pressures. The researchers obtained baseline Mach 0.8 fuel flow and performance data at altitudes of 35,000 to 50,000 feet.[299]

Project Bee

Following the 1949 reorganization, the Lewis lab had begun working with high-energy propellants such as diborane, pentaborane, and hydrogen. The potential was great, but the fuels were difficult to handle and required large tanks. The military became interested in the potential of these high-energy fuels, however. In late 1954, Lewis researchers studied the combustion characteristics of gaseous hydrogen in a turbojet combustor. Despite poor mixing of the fuel and air, it was found that the hydrogen provided over 90-percent efficiency.[300]

Almost immediately thereafter, Associate Director Abe Silverstein became focused on the possibilities of hydrogen. According to a 1974 interview with John Sloop, Silverstein realized that high-altitude aircraft would require a large fuselage and wingspans. These areas could be used for the large tanks required to store hydrogen.[301] He enlisted Eldon Hall to continue working on mixture ratios and then refine the calculations. They coauthored a report, not formally issued until April 1955, which foretold of liquid-hydrogen performing missions that far surpassed anything that traditional hydrocarbon fuels could perform.[302]

That fall, Silverstein secured a contract to work with the air force to examine the practicality of liquid hydrogen aircraft. The Lewis portion of the program was referred to as "Project Bee."[303] A new B-57B aircraft was obtained by the air force especially for this project, and a liquid hydrogen production plant,

Image 139: On the right: Abe Silverstein leads a tour of military and government officials in November 1955. At the time, Silverstein was negotiating with the air force to study the feasibility of a hydrogen-powered aircraft. An agreement was signed in December for a $1 million, one-year crash study. The program was called Project Bee. (NASA C-1955-40498)

the nation's first commercial facility for liquid hydrogen, was built in nearby Painesville, Ohio. The aircraft was powered by two Wright J65 engines and equipped with two 23-foot-long wing tanks, one of which was modified so that it could be operated using either traditional or liquid hydrogen pro-pellants. The other tank would be used to store helium used to pump the hydrogen.[304]

A number of tests were conducted on these engines in the AWT and the Four Burner Area altitude cells. Unlike prior studies in the AWT, which used external air to create flow, this test used internal tunnel air. Lower pressures were attained since the exhausters only had to make up for tunnel leakage, not leakage plus air exhausted out of the air scoop. In addition, an exhaust diffuser was used rather than the usual nozzle. With nozzles regulating the exhaust airflow, the tunnel pressure was less than half of the turbine's total

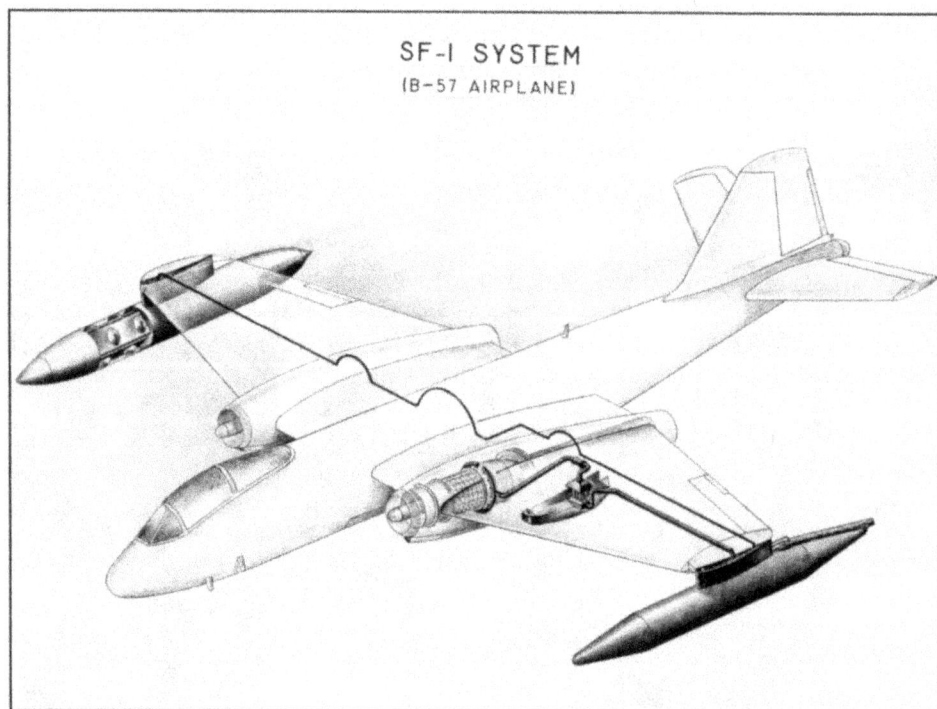

SF-I SYSTEM
(B-57 AIRPLANE)

Image 140: In 1957 a U.S. Air Force B-57B aircraft was brought to NACA Lewis for a series of liquid-hydrogen test flights. The aircraft was equipped with 23-foot-long, 430-gallon wing tanks. One of the tanks was modified to store liquid-hydrogen propellant and the other to store helium, which would be used to pump the hydrogen. The hydrogen tank consisted of a thin stainless steel inner shell covered with a lightweight, low-thermal-conductivity foam plastic insulation and an outer fiberglass covering to enclose the insulation and protect it against air erosion. (NASA C-1957-44419)

pressure. The diffuser permitted the tunnel pressure to be almost the same as the turbine pressure. The result was a 25,000 to 30,000 foot increase in altitude over previous AWT tests.[305]

An initial AWT study led by William Fleming compared the performance of liquid hydrogen with the standard JP-4 fuel in similar engines. The results clearly showed the combustion superiority of hydrogen. The JP-4 combustion became unstable above an altitude of 60,000 feet and failed at 80,000 feet, but hydrogen performed well at altitudes up to 90,000 feet.[306]

In a follow-up study in the AWT, Harold Kaufman sought to check the actual engine and fuel system that would be used on the B-57B test flights. A Wright J65-B-3 engine was run in both jet fuel and hydrogen modes. Again it was shown that the hydrogen outperformed the JP-4, particularly at high altitudes. Lewis researchers tested the switching between the jet fuel and hydrogen tanks numerous times with satisfactory results.

Image 141: A Wright J65-B-3 engine being prepared for runs in the AWT. After World War II, Wright Aeronautical did not invest heavily in the turbojet. Instead they concentrated their efforts on replacement parts and repairs of the existing Wright reciprocating engines. The J65, based on a Rolls-Royce Sapphire design, did not sell well. Wright became a subcontractor for companies that were previously their rivals.[307] (NASA C-1955-38288)

Walter Olsen, Head of the Fuels and Combustion Division, felt that they had proven the ability of the hydrogen system with these extensive ground tests, but Silverstein insisted on a flight test.[308] The intention was to take off using jet fuel, switch to liquid hydrogen while over Lake Erie, then after burning the hydrogen supply, switch back to jet fuel for the landing. Several dry runs were flown in the fall of 1956, and the first attempt at hydrogen-powered flight was attempted on 23 December 1956. The first two flights failed to make the switch to liquid hydrogen, but the third attempt, in February 1957, was a success.[309]

Merritt Preston, a research engineer at the time, recalled that the last flight occurred during a snowstorm. "[Abe] decided we needed to fly. We were obedient, so we flew." When it came time to switch over to the hydrogen fuel, the engine would not cooperate so the pilot had to return with a full load of hydrogen during inclement conditions." Landing with a full load of the highly explosive hydrogen was a daunting task. Preston added, "As if that were not enough, it was discovered that the landing gear would not deploy. The pilot circled until he was just about out of regular jet fuel when finally the landing gear lowered allowing the aircraft to land safely.[310] The flights would later be used to help convince NASA leadership that liquid hydrogen was safe to use for second-stage rockets.

Image 142: The B-57B aircraft flew several flights in the winter of 1957 using liquid hydrogen as a propellant at cruising altitudes. The flights would be one of the first times all of the lab's research elements were incorporated into a single operating system. This photograph of the aircraft was taken one year later with the wing tanks removed. (NASA C-1958-47236)

Getting Jet Engines Ready for Civilians

By the mid-1950s, the aircraft industry was close to introducing jet airliners to the nation's airways. The noise produced by the large jet engines, however, would pose a considerable problem for communities near airports. This problem was demonstrated by an NACA Lewis researcher who played long-play (LP) audio records of military jet engines for an audience at the NACA Lewis 1957 Inspection.[311]

The NACA had formed a Special Subcommittee on Aircraft Noise to coordinate research on the problem. AWT tests showed that the source of the loudest noise was not the engine itself, but the mixing of the engine's exhaust with the surrounding air in the atmosphere. The pressures resulting from this turbulence produced sound waves. Lewis undertook a variety of noise-reduction studies during the ensuing years involving engine design, throttling procedures, and suppressors.

One of the first studies sought to design an exhaust nozzle that reduced the turbulence. A number of nozzle configurations, including several multi-exit "organ pipe" designs, were created. The goal of the nozzles was to mix the

Image 143: The NACA Special Subcommittee on Aircraft Noise, chaired by William Littlewood, views a Pratt & Whitney J57 engine with a Greatex No. 2 exhaust nozzle in the AWT test section. The engine was tested with various nozzles as part of a noise-reduction study in the tunnel from January to May 1957. (NASA C-1957-44377)

exhaust with the surrounding air to slow down its velocity and decrease the noise. A Pratt & Whitney J57 was tested in the AWT with many of these nozzle configurations from January to May 1957.[312] It was found that the various nozzle types did reduce the noise levels, but they also reduced the aircraft's thrust. By the fall of 1957, however, it was announced that the addition of an NACA-developed ejector reduced the noise levels without diminishing thrust. The ejector did result in some drag and would thus limit the aircraft's range.[313]

Image 144: Various nozzles tried to suppress noise on the Pratt & Whitney J57 engine in the AWT. Although the nozzles were successful in reducing the turbulence and thus the noise, they also reduced the engine's thrust. (Left to right, then top to bottom: NASA C-1957-44564, C-1957-44819, C-1957-44710, C-1957-44227, C-1957-44737, and C-1957-44278)

Image 145: A J57 engine is tested with a Coanda nozzle and ejector for a noise-reduction study conducted in the AWT. As the engine exhaust roared through the ejector it pulled low-speed air from the ejector pump with it. This reduced the shock and resulting noise as the exhaust hit the atmosphere. (NASA C–1957–44771)

Image 146: A B-47 bomber on display at the NACA Lewis 1957 Inspection. One of its GE TG-190 (J47) engines has a noise-reducing ejector installed on its exhaust. The ejector was developed at Lewis and tested on a Pratt & Whitney J57 in the AWT. (NASA C–1957–46143)

Where Do We Go From Here?

By the mid-1950s aircraft engine research began plateauing. Benjamin Pinkel later recalled, "They were coming to a state of maturity where the results of research were very incremental. As a matter of fact, there was the feeling developing of, 'Where do we go from here?'" The answer came on 4 October 1957 with the launch of Sputnik I. The NACA and Lewis in particular had delved into rocket and high-energy fuels research, but it was mostly kept hidden from Congress and the public. The Soviets forced their hand, however, and Pinkel added, "[The space program] saved the day for the NACA."[314] Immediately following the Sputnik launch, work began modifying the AWT so that it could test rocket engines in altitude conditions.[315] The engines were relatively small, and the use of the AWT for rockets was relatively brief. The days of the AWT's role as a wind tunnel drew to an abrupt end.

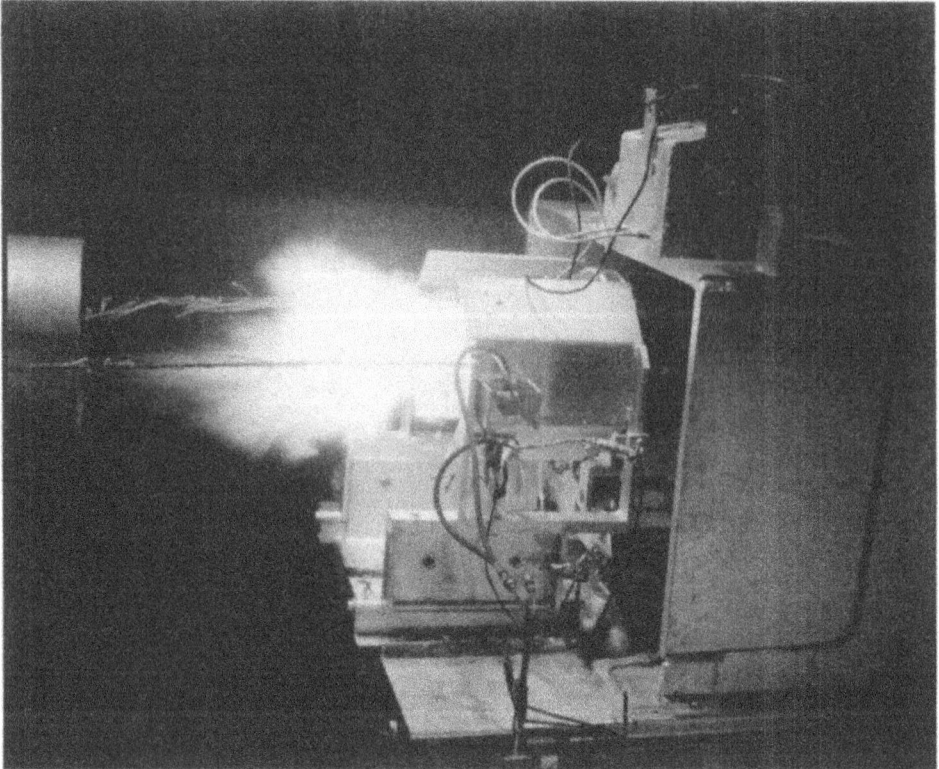

Image 147: Atlantic Research's grain Arcite rocket fires in the AWT. The tests ran from the fall of 1958 to early 1959. (NASA C-1959-49412)

Image 148: Solid-rocket assembly damage in the AWT. Unlike the solid propellants used for typical missiles or other engines, the behavior of propellants in high-performance vehicles cannot be predicted on the basis of sea-level tests. In May 1958, Lewis researchers commenced a series of low-pressure tests in the AWT to study the behavior of the solid propellant, high-performance rocket engines with high expansion ratios. The pressure's influence on the entire chamber, the exhaust nozzle, and the rocket's specific impulse was measured.[316] (NASA C–1958–48161)

Endnotes for Chapter 6

274. U.S. Army Center of Military History, "The Korean War: The Outbreak" (13 September 2006): 1-3, *http://www.history.army.mil/brochures/ KW-Outbreak/outbreak.htm* (accessed 27 August 2009).

275. Interview with Abe Silverstein, conducted by John Mauer, 10 March 1989, NASA Glenn History Collection, Cleveland, OH.

276. Abe Silverstein to staff, "Reorganization of Research Divisions," 21 December 1949, Directors' Collection, NASA Glenn History Office, Cleveland, OH.

277. "Director's Office Organizational Chart," 1955, NASA Glenn History Collection, Cleveland, OH.

278. U.S. Air Force, "General Electric J47 Turbojet Engine," National Museum of USAF, *http://www.nationalmuseum.af.mil/factsheets/factsheet.asp?id=873,* (accessed 27 August 2009).

279. "Turbojet Engine Development Budget Chart" 1955, C–1955–39607, NASA Glenn Imaging Technology Center, Cleveland, OH.

280. "mod_awt_c1 through mod_awt_c13" (1951), NASA Glenn Images, NASA Glenn Imaging Technology Center, Cleveland, OH.

281. Eugene Emme, *Aeronautics and Astronautics: An American Chronology of Science and Technology in the Exploration of Space, 1915–1960* (Washington, DC: NASA Headquarters, 1961), pp. 63–77.

282. Robert F. Dorr, "Engine Faults Dashed Demon's Navy Career," Army Times, (Springfield, VA, Army Times Publishing Co., 23 January 2006) *http://www. armytimes.com/legacy/rar/1-292308-1490368.php* (accessed 27 August 2009).

283. Harold Finger, Robert Essig, and E. William Conrad, *Performance of Rotor and Stator Blade Modifications on Surge Performance of an 11-Stage Axial Flow Compressor: I—Original Production Compressor of XJ40–WE–6 Engine* (Washington, DC: NACA RM–E52G03, 1953).

284. Adam Sobolewski and Robert Lubick, *Altitude Operational Characteristics of Prototype J40–WE–8 Turbojet Engine* (Washington, DC: NACA RM–E52L29, 1952).

285. Joseph Sivo and William Jones, *Operational Characteristics of RA–14 Avon Turbojet Engine* (Washington, DC: NACA RM–E56D09, 1956).

286. Carl Campbell and Henry Welna, *Preliminary Evaluation of Turbine Performance With Variable-Area Turbine Nozzles in a Turbojet Engine* (Washington, DC: NACA RM–E52J20, 1953).

287. Dorr, "Engine Faults Dashed Demon's Navy Career."

288. "Apprentices Learn New Skills," *Lewis News* (29 August 1969), p. 1.

289. Interview with Bill Harrison, conducted by Bob Arrighi, 14 October 2005, NASA Glenn History Collection, Cleveland, OH.

290. *Apprentice Manual* (Washington, DC: NACA Flight Propulsion Laboratory, 1953), p. 7.

291. "Apprentices Learn New Skills."

292. "Victory Predicts Great Future for Graduates," *Wing Tips* (15 February 1952).

293. Interview with Bob Walker, conducted by Bob Arrighi, 2 August 2005, NASA Glenn History Collection, Cleveland, OH.

294. *Apprentice Manual,* p. 13.

295. "Youths Complete Training in Pre-Apprentice Program," *Lewis News* (1 September 1967), p. 1.

296. Jeremy R. Kinney, "Starting From Scratch?: The American Aero Engine Industry, the Air Force, and the Jet, 1940–1960" (Reston, VA: American Institute of Aeronautics and Astronautics, 2003).

297. "J57 (JT3)," Pratt & Whitney, A United Technologies Company, *http://www. pw.utc.com/About+Us/Classic+Engines/J57 (JT3)/* (accessed 17 November 2007).

298. Robert Lubick, William Meyer, and Lewis Wallner, *Preliminary Data on the Effects of Altitude and Inlet Pressure Distortions on Steady-State and Surge Fuel Flow on the J57–P–1 Turbojet Engine* (Washington, DC: NACA RM–E55A06, 1955), p. 1.

299. Lubick, *Preliminary Data.*

300. V.F. Hlavin, E.R. Jonash, and A.L. Smith, *Low Pressure Performance of a Tubular Combustor With Gaseous Hydrogen* (Washington, DC: NACA RM–E54L30A, 1955), pp. 9–10.

301. John Sloop, *Liquid Hydrogen as Propulsion Fuel,* 1945–59 (Washington, DC: NASA SP–4404, 1978), chap. 6.

302. Abe Silverstein and Eldon Hall, *Liquid Hydrogen as a Jet Fuel for High-Altitude Aircraft* (Washington, DC: NACA RM–E55C28a, 1955).

303. Sloop, *Liquid Hydrogen.*

304. R. Mulholland, Joe Algranti, and Ed Gough, Jr., *Hydrogen for Turbojet and Ramjet Powered Flight* (Washington, DC: NACA RM–E57D23, 1957).

305. Harold R. Kaufman, *High-Altitude Performance Investigation of J65–13–3 Turbojet Engine With Both JP–4 and Gaseous-Hydrogen Fuels* (Washington, DC: NACA RM–E57A11, 1957), p. 2–3.

306. William A. Fleming, et al., *Turbojet Performance and Operation at High Altitudes With Hydrogen and JP–4 Fuels* (Washington, DC: NACA RM–E56E14, 1956), p. 18.

307. Kinney, "Starting from Scratch?"

308. Kaufman, *High-Altitude Performance,* p. 1–2.

309. Sloop, *Liquid Hydrogen.*

310. Interview with G. Merritt Preston, conducted by Carol Butler, 1 February 2000, Johnson Space Center Oral History Project, Houston, TX.

311. "Jet Aircraft Noise Reduction," Annual Inspection at NACA Lewis Flight Propulsion Laboratory, 7–10 October 1957, NASA Glenn History Collection, Cleveland, OH.

312. "Jet Aircraft Noise Reduction."

313. J.S. Butz, Jr., "NACA Studies Ways to Soften Jet Noise," *Aviation Week* (4 November 1957).

314. Interview with Benjamin Pinkel, conducted by Virginia Dawson, 4 August 1985, NASA Glenn History Collection, Oral History Collection, Cleveland, OH.

315. William A. Fleming to Virginia Dawson, "Summary of Experience at NACA Lewis Research Center Cleveland, OH," NASA Glenn History Collection, Cleveland, OH.

316. Carl Ciepluch, *Performance of a Composite Solid Propellant at Simulated High Altitudes* (Washington, DC: NASA TM–X–95, 1959), p. 1.

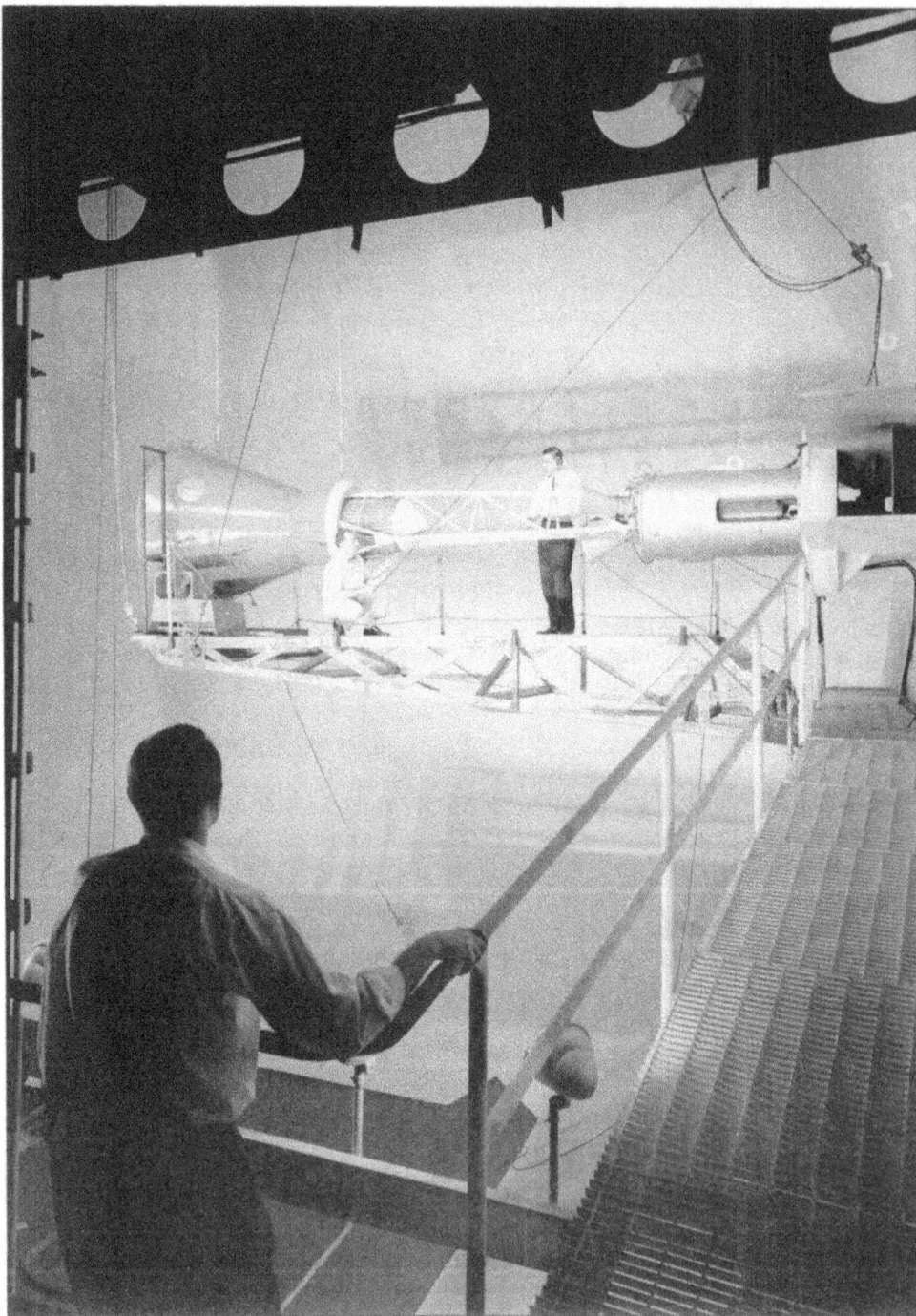

Image 149: A Mercury capsule is mounted in the Altitude Wind Tunnel for a test of its escape tower rockets. The AWT was quickly modified and became the center of NASA Lewis's work on Project Mercury. (NASA C-1960-53287)

Chapter 7

Space Is the Place | Project Mercury (1959–1960)

The deafening hiss of the control jets and the piercing alarm subsided. A former navy test pilot was helped out of the training rig, and threw up on the floor as NASA Lewis engineers guided him to a nearby cot. It would be an hour before he could regain his composure.[317] It was 16 February 1960, and Alan Sheppard, considered to be the finest pilot of the original astronaut corps, was the first of the Mercury 7 to attempt to operate the Multi-Axis Space Test Inertia Facility (MASTIF). The MASTIF was one of the highest profile test programs ever undertaken either in the Altitude Wind Tunnel (AWT) or at NASA Lewis Research Center. Over the previous two years, Lewis and the AWT had shifted focus from aeronautics to space, from the National Advisory Committee for Aeronautics (NACA) to the National Aeronautics and Space Administration (NASA), from research to development, and from a wind tunnel to an altitude chamber.

Public and congressional reaction to the 4 October 1957 *Sputnik I* launch by the Soviet Union demanded that the United States undertake a space program. The military, which had been attempting to launch its own satellite into orbit, was eager to lead the way in space, but President Dwight Eisenhower was steadfast in his desire to have a civilian agency direct the program. The NACA seemed to be a logical group on which to build the new space agency, but the NACA's Langley and Ames aeronautical laboratories were hesitant. It was only the Lewis lab in Cleveland that felt passionately about getting into space.[318] Over the next few years, the AWT would be the center of Lewis's contributions to the first human space mission, Project Mercury.

Image 150: Lewis's hangar was repainted after the NACA Lewis Flight Propulsion Laboratory was renamed the "NASA Lewis Research Center." Lewis, which had been investigating rocket engines and high-energy propellants for years, saw space as an extension of its aeronautics research. A host of new facilities were built for the space program, while others like the AWT were modified. (NASA C–1958–48854)

Lewis Guides the New Space Agency

NACA Lewis researchers had been advocating further space research for years. Predating *Sputnik* by two and one-half years, one document, "Suggested Policy and Course of Action for NACA Re Rocket Engine Propulsion," urged the NACA to enter the field of rocket engines. It stated that spaceflight "is a logical extension of the continuing success of NACA in advancing the engine art."[319] A second document, "Lewis Laboratory Opinion of Future Policy and Course of Action for the NACA," claimed that space exploration was imperative for the nation's survival during the Cold War. It called for an annual 25 percent increase in the NACA's staff, a new space laboratory, a launching center, communications center, and other facilities. The document served as an outline for the future NASA. It stated that the greatest need was for basic research under the aegis of a national agency independent from the military.[320]

Abe Silverstein was transferred to NACA Headquarters in early 1958 to assist NACA Director Hugh Dryden with the formation of the agency. Silverstein played a critical role in the creation of NASA. An appropriations bill, now referred to as "the Space Act," was submitted to Congress on 2 April to fund the new agency. NASA officially came into being on 1 October, and the NACA Lewis Flight Propulsion Laboratory became the NASA Lewis Research Center. Lewis underwent a major reorganization and for the next 10 years concentrated its efforts almost exclusively on the space program.

Silverstein was named Director of Space Flight Operations later in 1958. He was responsible for coordinating, supervising, and reviewing all spaceflight work done at the three NACA labs. He also managed the personnel and budget decisions for the Space Task Group (STG), whose assignment was to design and manage the nation's new space program. During his tenure at the NACA/NASA Headquarters, Silverstein not only laid out the framework for NASA and the manned space program but also planned early weather and communications satellite missions and numerous unpiloted missions, including Ranger, Mariner, Surveyor, Syncom, and Voyager. It was Silverstein who named NASA's first human spaceflight programs. Mercury represented "the swift messenger of things to come," and Apollo showed a progression, with Apollo being the greatest of the gods.[321]

Image 151: Left to right: Associate Administrator Robert Seamans, Deputy Administrator Hugh Dryden, Administrator James Webb, and Abe Silverstein at a press conference following Yuri Gagarin's 1961 orbital flight. Silverstein spent three years at NASA Headquarters helping create and guide the new space agency. (NASA GPN–2002–000153)

NASA's first effort would be Project Mercury—a series of 21 unmanned and 7 single-person orbital flights. NASA's Source Selection Board, chaired by Silverstein, selected McDonnell to manufacture the final *Mercury* capsules.[322] The boilerplate capsules would be designed and created at Langley and Lewis. Although Lewis was steeped in propulsion research, the STG felt that *Mercury's* schedule did not provide enough time for NASA to develop its own rockets. Instead, the group decided to launch the capsules on the military's existing Redstone and Atlas missiles.

The Altitude Wind Tunnel Applies Itself to the Space Program

Wind tunnels of all sizes and speeds were used for Project Mercury. The AWT was a significant facet in this new program, only not as a wind tunnel. It was the AWT's ability to simulate high altitudes inside a large chamber that piqued the STG's interest. The STG had originally wanted to qualify the *Mercury* capsule's retrorockets and instrumentation with a series of balloon drops near the edge of the atmosphere. By 22 May 1959 the STG decided to forgo the balloon drops and use the AWT instead.

Image 152: Harold Gold and Merritt Preston discuss NASA Lewis's role in Project Mercury on Dorothy Fuldheim's local Cleveland television show, The One O'Clock Club. Preston headed Lewis's STG branch, and Gold designed the guidance system for the Big Joe capsule. It was Lewis's first high-profile assignment in the national space program. (1959) (NASA C-2008-00902)

The turning vanes, makeup air pipes, and cooling coils were removed from the wide western end of the tunnel, and a device was installed inside the tunnel to test the capsule's attitude control system. During a 15 October 1959 meeting to allocate future Project Mercury assignments, the STG decided to also make use of the AWT to test the Atlas separation system, calibrate the capsule's retrorockets, study the escape tower rocket plume, and train astronauts to bring a spinning capsule under control. The tunnel would permit the retrorockets to fire at conditions found at altitudes of 80,000 feet.[323]

Big Joe's Attitude Control

The initial phase of Project Mercury consisted of a series of uncrewed launches using the air force's Redstone and Atlas boosters and the Langley-designed Little Joe boosters. The first Atlas launch, referred to as "Big Joe" to differentiate it from the smaller Little Joe rockets, was a single attempt early in Project Mercury to use a full-scale Atlas booster and simulate the

Image 153: Construction of the MASTIF inside the AWT to test the autopilot and attitude control systems for the Big Joe Mercury capsule. The MASTIF was built near the tunnel's throat section, with stairs installed to reach the test section and observation platform. (NASA C-1959-50266)

reentry of a mock-up *Mercury* capsule without actually placing it in orbit. The mission was intended to assess the performance of the Atlas and the reliability of the capsule's attitude control system, beryllium heatshield, and recovery process.[324] The launch was crucial since the STG felt that the entire manned space program relied on the power of the Atlas rocket and the dependability of the heatshield.[325]

The overall design of Big Joe had been completed by December 1958, and soon thereafter project manager Aleck Bond assigned NASA Lewis the task of designing the electronic instrumentation and automatic stabilization system. Lewis also constructed the capsule's lower section, which contained a pressurized area with the electronics and two nitrogen tanks for the retro-rockets. Lewis technicians were responsible for assembling the entire capsule: the General Electric (GE) heatshield, NASA Langley afterbody and recovery canister, and Lewis electronics.[326]

Image 154: Installation of the attitude control system for the Big Joe capsule in the AWT shop area. Robert Miller (on the left), Lou Corpas, Frank Stenger, and Phil Ross created the MASTIF test rig, which was based on a design by Harold Gold and Edward Otto. The upper portion of this piece contained the control system; gyros, thrusters, nitrogen tanks, and electronics were located in the lower section. (NASA C–1959–50457)

Image 155: NASA Lewis researcher William Masica operates the MASTIF. The Big Joe naviga-tion system package is mounted in the center of three cages, which each spin in a different direction. Mounts for the tunnel's turning vanes, which had been removed prior to these tests, can be seen in the background along the ceiling. (NASA C-1959-51728)

Following separation, the capsule had to position itself with its heatshield down. The autopilot system kept the capsule at the correct attitude during reentry, controlled the damping and roll during the actual reentry, and supplied information on the stability of an uncontrolled vehicle by shutting itself off after a specified gravity load (G load) was met.[327] Harold Gold and John Sanders oversaw the Lewis team that designed, built, and tested the cap-sule's attitude controls. These consisted of eight cold-gas nitrogen jets, their four tanks, and the autopilot gyros and electronics in a pressurized compart-ment between the floor and the heatshield.[328]

A mock-up capsule was built to test the performance of the control system. At that point, the NACA in-house mentality remained alive despite the new agency's trending toward contracting work out. Lewis engineer, Robert Miller, recalled that during NASA's first years, he would "take these drawings and go to the shops and have these guys study them. Then we'd have a meeting and we'd discuss them and create enough interest and say, 'Well, you know, can we do this in-house?' And of course they created enough interest so they would do it in-house instead of shipping it out."[329]

In March 1959, engineers began assembling the MASTIF inside the AWT. The device could recreate any possible motion that the capsule might encounter during its flight.[330] The MASTIF had three cages nested inside one another, each of which spun independently in roll, pitch, or yaw. The goal was to determine if the autopilot system could rectify the capsule's attitude following separation. If the control system failed to work properly, the heatshield would be out of place and the spacecraft would burn up during reentry.[331] The AWT tests verified the control system and thrusters.

Afterward Lewis technicians assembled the actual Big Joe capsule in Lewis's hangar. On 9 June, it was loaded on an air force transport aircraft and flown to Cape Canaveral. Scott Simpkinson led a team of 45 test operations personnel from Lewis who followed the capsule to Florida and spent the ensuing months preparing it for launch. The launch took place in the early morning hours of 9 September 1959. Simpkinson's team became concerned when they realized that two exterior boosters had not yet jettisoned. Gold recalled afterward, "At T plus four minutes all eyes turned to the telemeter panel. Finally the meters on the panel deflected. The control system was on and our hopes were high for a successful flight."[332]

Though the Atlas 1-D booster failed to properly separate, which prevented the spacecraft from attaining its intended altitude and speed, the launch procedures went well, the heatshield functioned properly, attitude control was accomplished, reentry flight dynamics were studied, and recovery was successful. It was the longest and fastest reentry recorded to date.[333] Warren Plohr described the mission, "Despite the dirty curve thrown by Atlas, the capsule reentered without excessive heating, landing in the water in perfect condition. Telemetry equipment began relaying data after capsule separation as planned, giving us much-desired information."[334] As a result of the mission's accomplishments, a second Big Joe launch was cancelled.

Image 156: The Big Joe capsule mounted atop an Atlas 1-D booster in the hours prior to its 3:19 a.m. launch on 9 September 1959 from Cape Canaveral. It was the first Mercury launch involving a full-scale booster. (NASA GPN–2002–000045)

Lewis Learns to Launch

Scott Simpkinson managed over 100 NASA Lewis engineers and technicians on the Big Joe project. Following the tests in the AWT, approximately 45 Lewis personnel, including those who had worked in the AWT, traveled to Cape Canaveral in mid-June 1959 to prepare the capsule for its launch. Although the Air Force was running the launch, the NASA team was adamant about staying involved. The Lewis contingent was stationed in Hangar S, a large structure that had few amenities but many anecdotes. These included transporting the capsule on a mattress in the back of a pickup truck and using the janitor's closets as offices.[335]

The switch from aeronautics to space was difficult for some NACA veterans at first. All procedures had to be documented and followed. There were no midtest or midflight adjustments. G. Merritt Preston, former AWT researcher and head of the Lewis Space Task Group, explained, "Before when we were flying up at Lewis, we'd just go out and do it, and that's not acceptable when you've got all this kind of business." The Lewis team soon adapted to the precise step-by-step procedures. "Sometimes it gets a little annoying, it's so detailed," admitted Preston, "but that's the way it really has to be."[336]

Image 157: The Big Joe capsule is shipped from Cleveland to Cape Canaveral on 9 June 1959. The capsule was assembled at Lewis. Later, the Mercury capsules were created at McDonnell Douglas. (NASA C–1959–50892)

The Big Joe capsule recorded 139 measurements, which covered temperature, pressure, noise, speed, and the behavior of various systems and equipment.[337] Frank Bechtel, who installed much of the telemetry, recalled working through muggy, mosquito-laden nights in Hangar S attempting to get the instrumentation installed. "It has like a thousand thermocouples embedded in this heatshield...We couldn't get it to work. Here we find out who ever built the connector for all the thermocouples...they didn't know how to solder a thermocouple connector and there was like a thousand wires in that connector. We had to rebuild that connector and resolder the whole thing."[338]

After almost three months of preparation at Cape Canaveral, the capsule and booster were ready for launch in the early morning hours of 9 September 1959. Preston, Simpkinson, Henry Plohr, Harold Gold, Hap Johnson, and Jacob Moser joined NASA Langley Big Joe project engineer Aleck Bond in

Image 158: *NASA Lewis engineers, technicians, and managers who traveled to Cape Canaveral for the launch of the Big Joe Mercury capsule. (6 Sept. 1959) (NASA C–2009–02180)*

the blockhouse to oversee the launch. Others, like Frank Bechtel, monitored telemetry consoles during the 3-hour countdown. He recalled, "For the launch I manned one station. I had to watch one temperature light. There was a whole bunch of guys. That was my job, watch one temperature light. And if something went wrong I had to let them know, 'Don't light the fuse!'"[339]

Although there was a problem with the Atlas booster's staging, Big Joe's attitude control system, heatshield, and recovery process worked well. G. Merritt Preston said afterward, "The success of the mission proves the capability of Lewis people involved with the operation."[340]

Some of the Lewis team stayed on at Cape Canaveral, where there were opportunities for both technicians and managers. Scott Simpkinson would go on to run the testing programs for both the Gemini and Apollo Programs and manage flight safety on the shuttle program. Preston became Director of Launch Operations for Mercury and Gemini and Manager of the NASA Johnson Space Center's Florida operations. Others, however, had ties to Cleveland, such as new homes or children in school, which made transfer unappealing. In addition, the stress associated with launching spaceflight hardware on a tight schedule was intense, and many people burned out.

The Infamous Multi-Axis Space Test Inertia Facility (MASTIF)

The extensive training and medical testing of the *Mercury* astronauts has been widely documented. NASA researchers were not sure how humans would react to the space atmosphere, so they were overly cautious in their preparation for the *Mercury* flights. In 2007, astronaut Scott Carpenter explained that more was known about the physical and mental health of the Mercury 7 than about any group of humans ever.[341] One of the most publicized and feared of the tests was in the AWT's MASTIF.

The three-axis gimballing rig, originally designed to test the attitude control of the Big Joe capsule, was adapted to train the astronauts how to rein in a tumbling spacecraft. NASA engineers feared that the capsule could spin out of control once its thrusters were fired. They added a pilot's chair, a hand controller, and an instrument display to the MASTIF in order familiarize the astronauts with the sensations of an out-of-control spacecraft.

NASA Lewis researcher James Useller and pilot Joe Algranti perfected and calibrated the MASTIF throughout 1959. Algranti responded well to disorientation on any one axis, but along two or more axes, his response time was slower. The pilots were spun at 20 rpm and then later 50 and 70 rpm. A number of other pilots were brought in to help the engineers perfect the test rig.[342]

An astronaut was secured in a foam couch in the center of the rig. The rig then spun on three axes from 2 to 50 rotations per minute. The pilots were tested on each of the three axes individually, then all three simultaneously. The yellow inner cage spun the pilots head over heels, the green cage rolled the pilots sideways, and the outer red yaw cage spun them horizontally. The pilot was strapped in with only arms free to operate the communication and panic buttons and a stick that controlled the small nitrogen jets that ran the movement of the MASTIF.

The rig, which included directional and vertical gyros, a rate damper, an amplifier calibration unit, and an accelerometer switch, was intended to mimic the *Mercury* capsule control system. The astronaut could alter any of the attitude gyros during an actual flight by firing any of 12 hydrogen peroxide rockets mounted to the outside of the axis pins to gain control of the rig after it had been set in motion.[343] One witness said that the jets "sounded like the blast of a giant air hose on concrete."[344]

Engineer Frank Stenger felt guilty watching the pilots struggle, so he decided to try it himself one afternoon. He recalled, "So I got in it and spun up in

Image 159: The MASTIF, a three-axis rig with a pilot's chair mounted in the center, was installed in the AWT to train Project Mercury pilots to bring a spinning spacecraft under control. After being secured in a foam couch in the center of the rig, an astronaut would use small nitrogen gas thrusters to bring the MASTIF under control. There was an incident during which one of the valves was opened and the empty rig spun wildly until it fell apart. The MASTIF was reassembled using the spare set of parts, although a different style of mounting had to be designed.[345] *(NASA C–1959–51723)*

just one axis, just tumbled head over heel, I think it was. And I went up to the maximum speed, coasted to stop, somehow, I don't remember whether I did it or whether we just coasted. But I got out of the cage and I was green. I was green. I was sick. I don't know if I upchucked, but think I sat around 2 hours trying to recover. I didn't go home until about 7 o'clock, just sitting somewhere."[346]

A second aim of the tests was to study the behavior of the astronauts' eyes when subjected to the rotations. During the rapid movements associated with the MASTIF or high-speed flight, the pilot's eyeballs tend to take longer to alter their direction than the body so the pilot is constantly readjusting his eyes to focus on what is in front of him. This condition, known as nystagmus, often blurs the vision. Lewis researchers attached electrodes to an astronaut's eye and to a camera inside the MASTIF's cockpit to film them during the spins. Robert Miller recalled, "I could see the eyes flipping up and down and sideways."[347]

In February and March 1960, the seven Project Mercury astronauts and female pilot Jerrie Cobb traveled to Cleveland to train on the MASTIF. Warren North and a team of air force physicians were on hand to monitor their health. After being briefed by Algranti and Useller, the rider would climb into the rig and be secured in the chair. Stenger would then slowly set the MASTIF in motion. It was the astronaut's job to bring it under control.[348] Each individual was required to accumulate 4.5 to 5 hours of MASTIF time.

The tests were grueling. Alan Sheppard said at the time that the MASTIF had given him the "most realistic vertigo experience of his life."[349] Wally Schirra compared a ride in the rig to a dog chewing on one's leg. Scott Carpenter explained, "No matter how well you fit the couch and no matter how well you're strapped in it, your internal organs start to move around even at slow speeds. And at high speeds, it really starts sloshing."[350] While preparing for his 1999 STS-95 mission, John Glenn remarked, "When I went into training for this flight, I was glad that rig had been torn down."[351]

A crude dressing room, known as the "Astro-Penthouse," was temporarily constructed on the AWT viewing platform next the test section. It was equipped with a cot for astronauts to recuperate on after their rides. The astronauts would descend into the test section and walk through the tunnel's throat section to the waiting MASTIF.

The MASTIF tests determined that rotations up to 70 rpm had no measurable influence on an astronaut's operation of the rig. Repetition of these tasks reduced the error rate. It was also found that when the astronauts stared at a single location, the effects of nystagmus were reduced significantly. Despite Sheppard's nausea during those first runs, severe motion sickness was surprisingly rare during normal operation of the MASTIF.[352] Stenger said, "It was a confidence level. The astronauts knew that once they got in here. And every one of them was able to reorientate themselves for reentry in a position.... Once they built up their confidence level then they knew they could [reorientate themselves] if they were tumbling. It's something they didn't know they could do."[353]

The first week was assigned to Sheppard and Schirra. On those first runs Sheppard did well in bringing the machine under control on one and two axes. Once all three axes were in motion, however, Sheppard became ill and had to activate the "chicken switch" that signaled to the engineers to shut down the machine. Sheppard doggedly tried the MASTIF again that afternoon. The trials improved, but again he was forced to call off the test before he could bring it under control. Over the next few days, however, Sheppard continued to climb into the rig for additional runs. Soon, he found a method by which he would control the MASTIF one axis at a time until he could bring the machine to a standstill. He had become adept enough to rein in 30 rpm on each axis. Sheppard was the first of the astronauts to master the machine. He later felt that his motion sickness during the tests might have been an early sign of the Ménière's inner ear syndrome that caused him to be removed from flight status between 1964 and 1969.

During the first U.S. suborbital flight, Sheppard assumed control of the capsule for a significant portion of the time outside the atmosphere. The hand controller, identical to the one he had mastered during his MASTIF runs, was used to pitch the capsule in various positions as it streaked into space at 5,100 mph.[354]

Photo Essay 4:
The Mercury 7 Come to Town

Image 160: Left to right: Robert Miller, John Glenn, and James Useller. The men were the driving forces behind the creation of the MASTIF and the running of the tests. (NASA C–1960–52744)

Image 161: Center left and right: Astronaut Alan Sheppard and NASA Lewis Research Center test pilot Joe Algranti discuss the MASTIF inside the AWT. Sheppard was the first of the Mercury astronauts to operate the rig. (NASA C-1960-52706)

Image 162: John Glenn begins his tests on the MASTIF. (NASA C-1960-52741)

Image 163: Left to right: Astronauts Deke Slayton and Scott Carpenter talk outside the "Astro-Penthouse." When asked about the MASTIF's affect on his eyes, Slayton said, "To clear up the vision, all the spaceman has to do is stop using the control forces of the nitrogen gas." (NASA C–1960–52807)

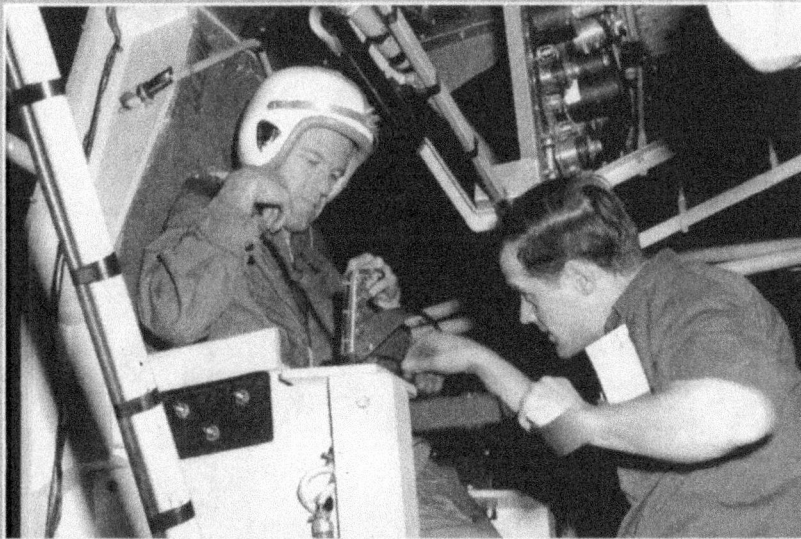

Image 164: Gordon Cooper prepares for a spin in the MASTIF. Cooper arrived in Cleveland after the others had finished their tests because of an illness in the family. (NASA C–1960–52972)

Image 165: Left to right: Gus Grissom and Wally Schirra pose next to the MASTIF inside the AWT. (NASA C-1960-52465 and C-1960-52818)

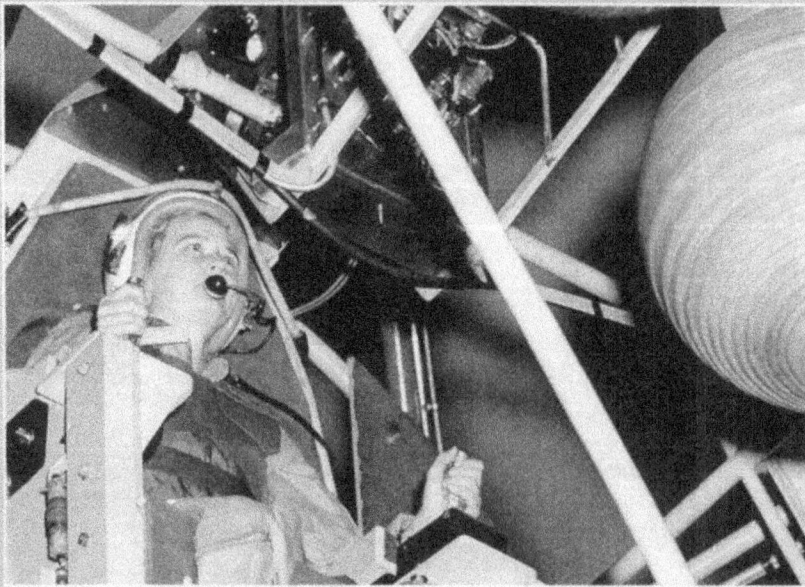

Image 166: Female pilot Jerrie Cobb prepares for a test on the MASTIF. Cobb was one of several female pilots who underwent skill and endurance testing similar to that of the Project Mercury astronauts. Several nonastronaut test pilots took turns on the MASTIF, but Cobb was the only woman. The NASA requirement that astronauts be military fighter pilots automatically disqualified potential female candidates. (NASA C-1960-53087)

Verifying the Retrorockets

The *Mercury* capsule had six rockets on a "retropackage" affixed to the bottom of the capsule. Three of these were posigrade rockets used to separate the capsule from the booster and three were retrograde rockets used to slow the capsule for reentry into the Earth's atmosphere. Performance of the retrorockets was vital since there was no backup if the systems failed. The Space Task Group (STG) assigned the NASA Lewis and NASA Ames research centers the task of verifying their reliability.[355]

NASA engineers feared that the posigrades would damage the booster during the separation process. During early 1961, separation tests for both the Atlas and Redstone boosters were conducted inside the AWT at altitudes comparable to the upper atmosphere. The setup was similar for both series of tests. The full-size capsule model was affixed to the horizontally mounted booster mock-up. The capsule would fire its posigrades and swing away from the booster on a pendulumlike tether.

The Redstone tests revealed that the gas that filled the booster's ballast section when the posigrades were fired actually helped the separation process by driving the capsule away from the booster at an additional 25 feet per second.[356] The ensuing *Mercury*-Atlas separation tests ensured that the firing of the posigrade rockets did not injure the booster rocket, retrorockets, or electronics. It also determined the actual boost level of the posigrades.[357] Although all three retrorockets fired simultaneously, only one was needed to achieve proper separation.[358]

Three retrograde rockets, also located on the bottom of the capsule, served as a braking system to be initiated approximately 3,000 nautical miles before the splashdown target. The three rockets were fired rotationally every 5 seconds for 1 minute before the entire retropackage was jettisoned.[359] A thrust stand was set up in the AWT to test the rockets. The exact position of the retrorockets' thrust was determined by using a thrust cell and two torque cells.[360] The studies showed that a previous probem with delayed ignition of the propellant had been resolved. Follow-up test runs verified the reliability of the igniter.[361] Lewis researchers were able to calibrate the retrorockets so that they would fire through the capsule's center of gravity and not send the spacecraft tumbling.[362]

Image 167: Mercury astronaut Scott Carpenter studies the separation system on a mock-up Mercury capsule in the AWT. The tunnel was used to study the posigrade rockets that separated the capsule from its booster after leaving Earth's atmosphere. (NASA C-1960-52808)

Image 168: Mercury capsule/Redstone separation test in the AWT. The posigrade rockets were fired, and the capsule jettisoned forward on a tether. High-speed film was used to analyze the separations. A plywood shield was hung guillotine-style above the ring. Once the bolts blew and the capsule shot out, the plywood dropped down to protect the booster model when the capsule swung back on its pendulum.[363] (NASA C-1960-52778)

Image 169: Left: Gale Butler examines the Mercury capsule's retrograde rockets prior to a test run inside the AWT. These studies showed that a previous problem concerning delays igniting the propellant had been resolved. Follow-up runs verified the reliability of the coated igniter's attachment to the propellant grain. In addition, the capsule's retrorockets were calibrated so that they would not alter the capsule's position when fired. (NASA C-1960-53146)

Image 170: From March to July 1960 the AWT was used to qualify escape tower rocket motors for the Mercury capsule. For the tests, the AWT's atmosphere was evacuated to an altitude of approximately 100,000 feet, and the escape tower was mounted to the tunnel wall with a mock-up Mercury capsule at the end. (NASA C-1960-53288)

Escape Tower Tests

The AWT was also used to determine if the smoke plume from the capsule's escape tower rockets would shroud or compromise the spacecraft. The escape tower—a 10-foot steel rig with three small rockets—was attached to the nose of the *Mercury* capsule. It could be used to jettison the astronaut and capsule to safety in the event of a launch vehicle malfunction on the pad or at any point prior to separation from the booster. The abort sequence could be initiated by ground control, the astronaut, or the booster's catastrophic detection system. Once actuated, the escape rockets located at the top of the steel tower would fire, and the capsule would be ejected away from the booster. After the capsule reached its apex of about 2,500 feet, the tower, heatshield, retropackage, and antenna would be ejected, and a drogue parachute would be released.[364]

Flight tests of the escape system were performed at Wallops Island as part of the series of Little Joe launches. Although the escape rocket fired prematurely on the first attempt in August 1959, the January 1960 follow-up was successful. Afterward, the AWT was used to qualify the escape rocket motors at altitudes of 100,000 feet. The escape tower was mounted to the tunnel wall with a mock-up *Mercury* capsule at the end. Three different escape motors were successfully fired. Thrust misalignment was studied on a fourth rocket using a four-piece balance system.[365]

The escape rockets were powerful enough to send a blast sailing through the entire wind tunnel. Sections of tunnel walls were painted white to assist with the high-speed photography. Every time the rockets were fired, however, they would coat the interior with soot and the walls would have to be cleaned. NASA Lewis engineer Lou Corpas remembered, "[The exhausters] pumped all this soot and stuff out, just dumped it out through their smokestacks and it covered everything, everybody's car, the street, every thing."[366]

Image 171: Astronaut Alan Shepard is seen on the deck of the USS Champlain after the recovery of his Mercury capsule. Shepard had just become the first American to enter outer space. (NASA S61–02727)

The Altitude Wind Tunnel Completes Another Mission

Once again, with little advance warning, the AWT had been reconfigured to contribute to a new national aerospace need. When the facility had come online just 15 years before, no one could have foreseen that it would be used as an altitude tank for manned space missions. The use of the AWT during the busy early years of Project Mercury sped up the preflight testing and saved the new agency money.[367] The attitude control and heatshield instrumentation on the Big Joe capsule verified that reentry was safe for humans. The escape system was never deployed during an actual launch, but the tumbling capsule scenario did manifest itself on Neil Armstrong and Dave Scott's Gemini VIII mission. By June 1960, almost a year before Alan Sheppard's suborbital flight, the AWT had wrapped up its first and only work on the human space program. It was ready for its next mission.

Endnotes for Chapter 7

317. Neal Thompson, *Light This Candle: The Life and Times of Alan Sheppard America's First Spaceman* (New York, NY: Crown Publishers, 2004), p. 187.
318. Interview with Abe Silverstein, conducted by John Sloop, 29 May 1974, NASA Glenn History Collection, Oral History Collection, Cleveland, OH.
319. "Suggested Policy and Course of Action for NACA Re Rocket Engine Propulsion," 6 May 1955, NASA Glenn History Collection, Cleveland, OH.
320. "Lewis Lab Opinion of Future Policy and Course of Action for the NACA," NASA Glenn History Collection, Cleveland, OH.
321. Arnold S. Levine, *Managing NASA in the Apollo Era* (Washington DC:, NASA SP-4102, 1982) chap. 2.
322. Levine, *Managing NASA,* chap. 2.
323. James Grimwood, *Project Mercury: A Chronology* (Washington, DC: NASA SP–4001, 1963), Part II-A. [Maxime A. Faget (Chief, Flight Systems Division) to Project Director, "Status of Test Work Being Conducted at the Lewis Research Center in Conjunction with Project Mercury," 22 October 1959.]
324. Space Task Group, "Project Mercury Preliminary Flight Test Results of the 'Big Joe,' Mercury R and D Capsule," NASA Project Mercury Working Paper No. 107 and NASA TM–X–73017, 1959.
325. Lloyd Swenson, James Grimwood, and Charles Alexander, *This New Ocean: A History of Project Mercury* (Washington, DC: NASA SP–4201, 1966), chap. 9.
326. "Big Joe, Lewis' Part in the Project Mercury Story," *Orbit* (22 May 1959): 3.
327. Swenson et al., This New Ocean.
328. "Big Joe, Lewis' Part," *Orbit*.
329. Interview with Lou Corpas, Frank Stenger, and Robert Miller, conducted by Bonita Smith, 17 September 2001, NASA Glenn History Collection, Cleveland, OH.
330. "Big Joe, Lewis' Part," *Orbit*.

331. Corpas, Stenger, and Miller interview, conducted by Smith, 17 September 2001.
332. "It Was Like This at Canaveral," *Orbit* (25 September 1959): 1.
333. Swenson et al., *This New Ocean.*
334. "News Flash From Cape Canaveral," *Orbit* (11 September 1959): 1.
335. Swenson et al., *This New Ocean.*
336. Interview with G. Merritt Preston, conducted by Carol Butler, 1 February 2000, Johnson Space Center Oral History Project, Houston, TX.
337. Swenson et al., *This New Ocean.*
338. Interview with Frank Bechtel, conducted by Bob Arrighi, 19 July 2004, NASA Glenn History Collection, Cleveland, OH.
339. Bechtel interview, conducted by Arrighi, 19 July 2004.
340. "It Was Like This at Canaveral," *Orbit.*
341. Scott Carpenter, "Talk at NASA Glenn Research Center, Cleveland, OH, 14 September 2007."
342. Swenson et al., *This New Ocean,* chap. 8.
343. "Astronauts at Lewis," *Orbit* (19 February 1960): 1.
344. Jean Pearson, "Thrills Await 7 Astronauts on Whirligig." *Detroit Free Press* (8 December 1959).
345. Corpas, Stenger, and Miller interview, conducted by Smith, 17 September 2001.
346. Corpas, Stenger, and Miller interview, conducted by Smith, 17 September 2001.
347. Corpas, Stenger, and Miller interview, conducted by Smith, 17 September 2001.
348. Pearson, "Thrills Await 7 Astronauts."
349. "Lewis Letter" (24 February 1960), Office of Public Information, NASA Glenn History Collection, Cleveland, OH.
350. Jean Pearson, "Project Dizziness," *Detroit Free Press* (6 March 1960), p. 20.
351. "Well Wishers Thank Glenn in a Big Way," *Aerospace Frontiers* (February 1999).
352. James Useller and Joseph Algranti, "Pilot Reaction to High Speed Rotation," *Aerospace Medicine* 34, no. 6 (June 1963).
353. Corpas, Stenger, and Miller interview, conducted by Smith, 17 September 2001.
354. Thompson, *Light This Candle,* p. 254.
355. Swenson et al., *This New Ocean.*
356. George C. Marshall Space Flight Center, *The Mercury-Redstone Project* (Washington, DC: NASA TM–X–53107, 1963), pp. 4–7.
357. Grimwood, *Project Mercury.* [Memo, Faget, 22 October 1959.]
358. McDonnell Aircraft Corporation, "Project Mercury Familiarization Manual," (St. Louis, MO: McDonnell SEDR–104, 1959), 7–3.
359. McDonnell Aircraft Corporation, "Project Mercury," 8–1.
360. Corpas, Stenger, and Miller interview, conducted by Smith, 17 September 2001.
361. James Grimwood, *Project Mercury.* A Chronology, Part II–A. [NASA Space Task Group, *Project Mercury [Quarterly] Status Report No. 6 for Period Ending April 30, 1960.]*
362. Grimwood, *Project Mercury.* [Memo, Faget, 22 October 1959.]
363. Corpas, Stenger, and Miller interview, conducted by Smith, 17 September 2001.
364. McDonnell Aircraft Corporation, "Project Mercury," 6–13, 5–12.
365. Grimwood, *Project Mercury:* [NASA Space Task Group].
366. Corpas, Stenger, and Miller interview, conducted by Smith, 17 September 2001.
367. Swenson et al., *This New Ocean,* chap. 6.

Image 172: Conversion of the Altitude Wind Tunnel into two altitude chambers. This photograph shows the largest of the three bulkheads being inserted into the tunnel. It is being placed approximately where the wind tunnel fan was located. (NASA C-1961-58551)

Chapter 8

Metamorphosis | Wind Tunnel to Vacuum Chamber (1961–1963)

After the final boom of the *Mercury* capsule escape rocket faded throughout the Altitude Wind Tunnel (AWT) in late May 1960, the facility's test queue was empty. During NASA's first two years of existence, the NASA Lewis Research Center had refocused its efforts almost completely on the space program. Less than 10 percent of the annual budget was dedicated to aeronautics.[368] The national human space program would continue, but much of the testing would take place at the NASA Langley Research Center and the Manned Space Flight Center being built in Houston. New facilities at Lewis, such as the Propulsion Systems Laboratory and Rocket Engine Test Facility, were better equipped than the AWT to test rocket engines.

In the aftermath that followed President Kennedy's April 1961 "Urgent Needs" address to Congress, NASA was given a seemingly unlimited budget. To accomplish the accelerated lunar landing mission, the Agency reorganized and began swelling its ranks through a massive recruiting effort. Lewis personnel increased from approximately 2,700 in 1961 to over 4,800 in 1966.[369] In October 1961 Abe Silverstein returned from NASA Headquarters to serve as the Director of the NASA Lewis Research Center.

Lewis also began a new wave of construction both in Cleveland and at its Plum Brook Station 60 miles to the west. Nineteen facilities were built or reassigned for space-related testing during the Apollo Program. Other aeronautics-based facilities, including the Small Supersonic Tunnels and the Icing Research Tunnel, were shut down.[370] It was in this atmosphere that NASA Lewis decided to create two large altitude chambers inside the AWT. It would prove to be a fortuitous decision. After acquiring the Centaur rocket program in 1962, Lewis would need the large chambers for a variety of tests.

Age of the Space Tank

The new and modified facilities were part of NASA Lewis's plan to create a comprehensive portfolio of test capabilities for a wide range of space applications. Former Centaur Manager, Larry Ross, explained that the new facilities "were more in line with the broad vision of a very ambitious space program, including nuclear propulsion. So [Silverstein's] vision, and his commitment to that broader ambitious program he saw the nation undertaking, led to the development of the facilities."[371]

One capability considered necessary by Lewis was a large space simulation chamber. Initial spaceflights revealed that the behavior of engines, operating systems, and flight hardware were affected by the environmental conditions encountered in space. Testing these components in simulated space settings would be vital to their success. This was particularly true of satellites, whose equilibrium temperatures were often linked directly to their survival.[372]

Silverstein was among the strongest advocates for extensive ground testing in simulated flight conditions. He had been a primary advocate of the space environmental simulation chamber built in late 1960 at McDonnell to fly simulated missions leading up to the first manned *Mercury* flights.[373] Silverstein believed that mission success could be guaranteed only if "every piece of equipment that is taken aboard…[is] environmentally checked so that it can live in the environment of space, the total environment of space: the vacuum of space, the temperatures of space, the pressures of space; that is, in the fields of other components; that the whole ensemble put together is tested environmentally so as to operate successfully."[374]

In early 1961 Lewis management decided to convert the AWT into two test chambers, one to simulate the vacuum of outer space and the other the conditions of the upper atmosphere. The project was estimated to cost $350,000 over two years.[375] The new Space Power Chambers (SPC) facility was part of the first wave of large space tanks in the United States. The AWT recreated the pressures, temperatures, and speeds of atmospheric flight, but recreating a space environment required much lower pressures, colder

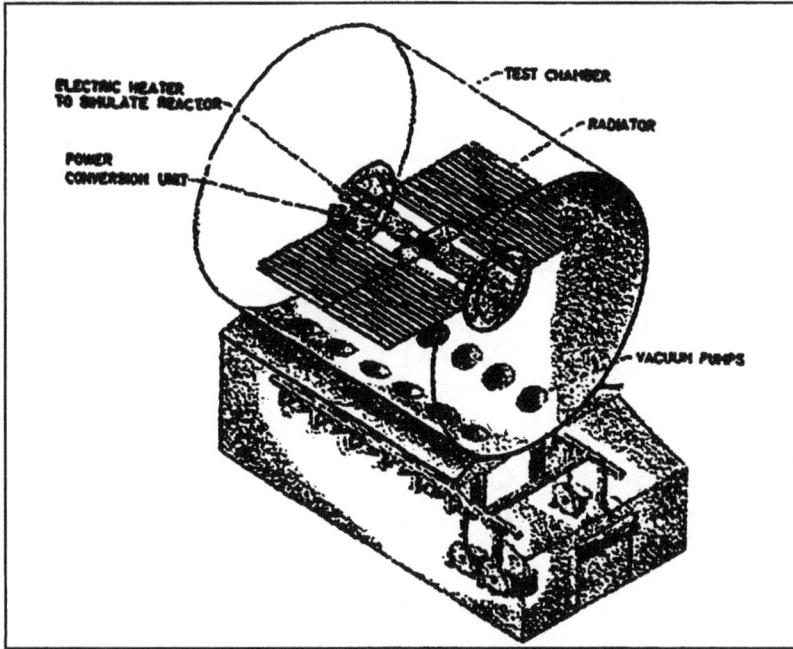

Image 173: Initial configuration of Space Power Chamber (SPC) No. 1 inside the AWT with a satellite installed. (NASA Glenn, Orbit, p. 1, 12 Jan. 1962)

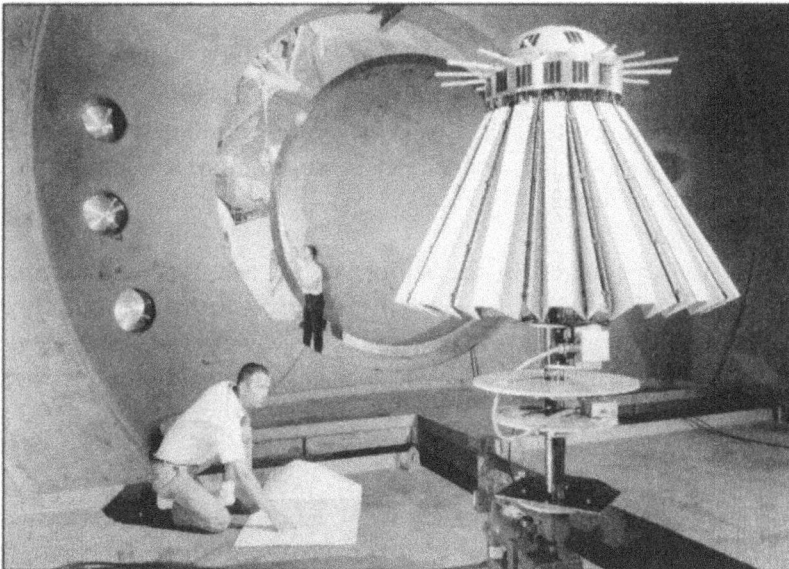

Image 174: Meteor Impact Model in the new SPC No. 1. This satellite was set up in the chamber days after the facility was officially rededicated as the Space Power Chambers (SPC) on 12 September 1962. The space tank was envisioned for the study of small satellites like this one. (18 Sept. 1962) (NASA C-1962-61717)

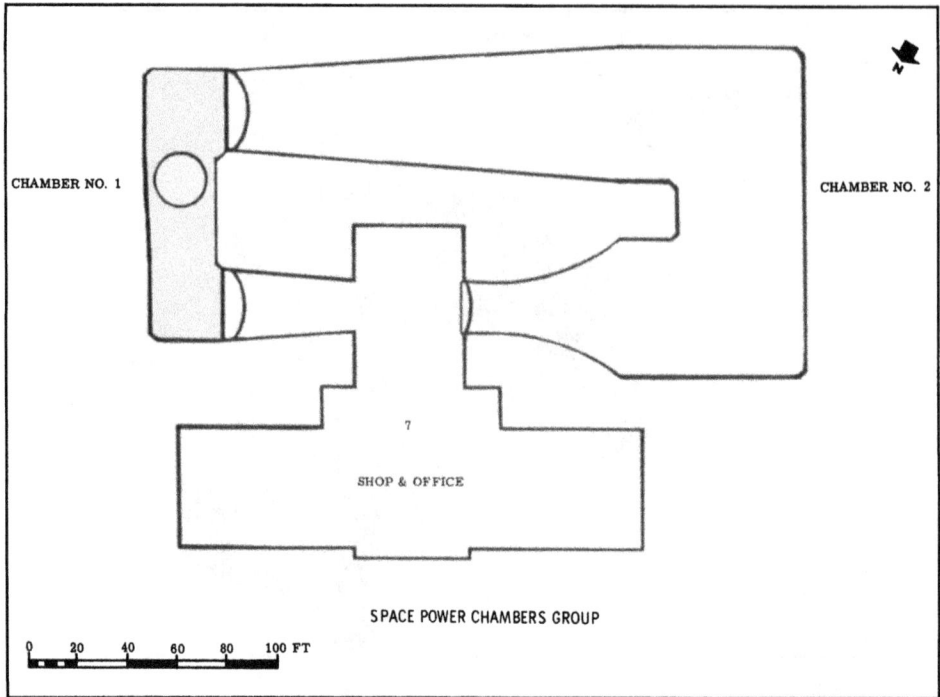

Image 175: SPC No. 1 vacuum chamber in relation to the overall SPC facility. (NASA Glenn)

temperatures, and a method of simulating solar radiation. Although several of the Project Mercury tests in the AWT were conducted in high-altitude conditions, it was not the space environment that satellites and space vehicles would encounter.

One study was conducted in the shop area of the AWT at this time—the Mercury Evaporating Condensing Analysis. Lewis researchers studied mercury condensation in an effort to design better power-conversion devices for the space program. Tests first were run in a ground facility then replicated in the air during microgravity flights on Lewis's AJ-2 aircraft.

There were no large vacuum chambers operating in the country when construction of the SPC began in the summer of 1961. The McDonnell tank was relatively small. By the end of 1962, however, there were 10; and by 1965, there were 40 including another chamber at Lewis for studying ion engines—the Electric Propulsion Laboratory. When it became operational in September 1963, the new 31-foot-diameter, 100-foot-long SPC No. 1 vacuum tank was rivaled in size only by the 35-foot-diameter, 65-foot-long Mark I

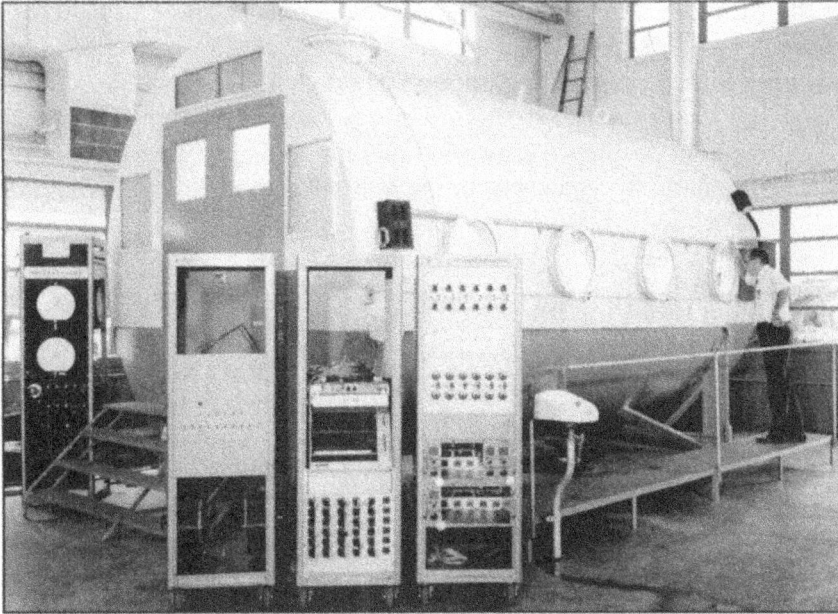

Image 176: Mercury Evaporating Condensing Analysis setup in the SPC shop area. It was the only research performed in the facility between March 1960 and September 1963. (NASA C–1961–57517)

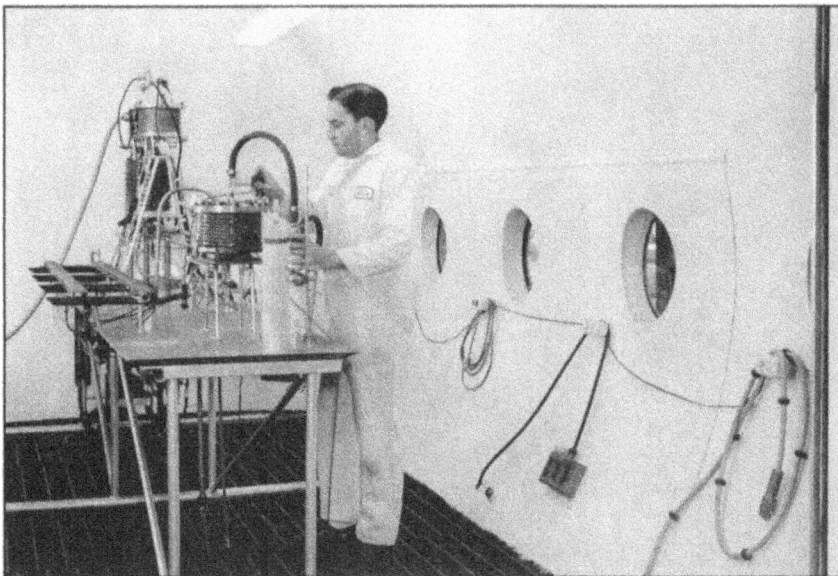

Image 177: Mercury was released into the horizontal pressurized condensing tube. Its behavior was monitored by an oscillograph and a high-speed camera on a rail that could film at any point along the tube.[376] Although the SNAP-8 program was eventually terminated, these power-conversion studies contributed to the Brayton cycle and the mercury Rankine systems of the late 1960s and early 1970s.[377] (C–1961–58584)

tank at the Arnold Engineering Development Center.[378] The Electric Propulsion Laboratory had several smaller vacuum tanks designed specifically for electric or ion thruster research. Although larger chambers capable of deeper vacuums would later be constructed, the rapid conversion of the AWT into the SPC allowed the facility to play a vital role in the early years of the space program, particularly in regards to the Surveyor missions to the Moon.

Reconfiguring a Giant for the Space Age

The AWT's 1-inch-thick shell, relatively smooth interior surfaces, strong infrastructure, and large internal volume were a good basis for a large vacuum chamber. The tunnel's drive fan, turning vanes, exhaust scoop, and test section lid were dismantled and removed. A section of the tunnel shell near the southeast corner was cut out and temporarily removed so that a

Image 178: A worker cuts away the AWT's outer shell in preparation for the installation of a large bulkhead and the sealing of the inner shell during the creation of the new SPC No.1 space tank. (NASA C–1961–58124)

large bulkhead could be placed inside the tunnel. Another seal, this one with a 15-foot-diameter access door, was placed near the tunnel's northeast corner, just beyond the test section. A third seal was made at the tunnel's contraction just before the test section to form the larger chamber in the west leg of the tunnel. It could be removed to bring test articles into the large SPC No. 2 test chamber.

A new pump house, built directly beneath SPC No. 1, housed an oil-diffusion-based vacuum system. The vacuum inside the chamber was brought down in three phases. The center's central exhaust system pulled the vacuum down to a pressure altitude of 100,000 feet in about 15 minutes, 2 mechanical roughing pumps in the pump house reduced the pressure further, and then the 10 diffusion pumps created the final vacuum pressures. The entire process required approximately 24 hours. Gaseous nitrogen could be introduced to rapidly return the chamber to sea level.[379]

The most frustrating aspect of the conversion process was the discovery that the tunnel's shell was not sealed sufficiently to maintain the desired vacuum level for the space tank. The seal had been sufficient for the wind tunnel's operation, but tests revealed that the entire shell would have to be rewelded for the vacuum chamber. Harold Friedman recalled, "[During] the war effort, it was hard to get welders. Just everything was wrong. In fact, I think the welds in the wind tunnel were barely passable."[380] Bill Harrison added, "And we ended up putting $25,000 into SPC and we rewelded every seam all the way around that big thing...in the process of doing that we found where the welders [in the 1940s] had laid welding rod in the crack and welded over top of it so that they could fill it up faster."[381]

The rewelding would not be required for the larger SPC No. 2 chamber at the other end of the tunnel. SPC No. 2 would be used for shroud separation tests in the same manner as it had been for the Project Mercury tests.[382] Howard Wine explains, "They didn't have to go up to near the altitude because the separation of the nose cone would take place when it was still in the atmosphere. So with the existing exhausters [from the AWT] were sufficient in order for them to get into the area that they could separate at some reasonable altitude."[383]

Contractors Get a Bad Name

One of the differences between NASA and the former NACA was the use of contractors to perform many of the tasks previously assigned to federal employees. The resulting resentment of the contractors by civil servants began in the early 1960s and persisted for decades. Several bad experiences during the construction of the SPC contributed to this impression of contractor inferiority.

One episode involved a man hired to sandblast the old paint from the vast interior of the SPC. The worker attempted to accomplish the task using an air gun that was fed from a small bucket of sand. The work proceeded at a slow pace, and eventually the contractor was replaced. Under the agreement, however, he had to clean up the sand and paint chips. After pulling his pickup truck under the hatch in the opposite end of the tunnel, he began shoveling the waste out by hand. This, too, proceeded slowly and the contractor was relieved of this duty, as well. NASA staff assumed the task and cleaned out the tunnel using large pieces of equipment.[384]

Another frustrating incident occurred during the assembly of the copper cold wall inside SPC No. 1. The cold wall contained a series of long, thin nitrogen tanks that had to be joined together. The assembly required a special type of arc welding know as heliarc welding. The crew selected to perform the work

Image 179. Workers inside SPC No. 1 work at removing the former tunnel's turning vanes and cleaning up the chamber's interior. (NASA C–1961–58128)

had no experience with this type of welding, so they hired a specialist from Pennsylvania and purchased new equipment. The crew practiced the technique in the parking lot for hours until they felt confident enough to begin the task.[365]

The final welds on the nitrogen bottles were not successful, however, and leakage from these cold-wall joints prevented the chamber from reaching the proper vacuum level. The contractor was let go, and NASA personnel took over the task. The nitrogen bottles were filled with helium, and small plastic bags were placed over each weld to check for leakage. Lewis retiree Howard Wine recalls receiving a mandate from Bill Harrison to complete the job rapidly. "We ran two shifts 24 hours a day, 12-hour shifts. I think it was six or seven days a week. We were against a time constraint." After several weeks of testing and rewelding, the chamber atmosphere was successfully reduced to the proper vacuum level.[386]

Those working on the Centaur Program would soon come to appreciate the role that contractors could play in the space program, but this opinion was not generally shared by the mechanical and technical staff at Lewis. The Apprentice Program was phased out in the 1970s, and the Test Installations Division and the Technical Services Division were merged. The use of outside contractors for this work increased, particularly in the early 1980s. As the years have gone by, however, the quality and performance of contractors has been proven and accepted as part of the NASA culture. In 2006 approximately half of the center's staff were contractors.[387]

Image 180: Construction of a new control room for SPC No. 1 beneath the tunnel's test section. (28 Nov. 1961) (NASA C–1961–58575)

Photo Essay 5:
The Altitude Wind Tunnel Is Transformed into the Space Power Chambers

Image 181: The AWT's outer steel shell is removed so that the inner steel tunnel can be checked for leaks. The middle layer of insulation is visible. It was found that, in the push to complete the original tunnel construction during World War II, the seams were poorly welded. To create the vacuum chamber, NASA would have to reweld the entire east end of the tunnel. (NASA C–1961–57394)

Image 182: Workers install a 31-foot bulkhead inside the southeast corner of the tunnel. The interior of the SPC was sandblasted and repainted with a double coat of aluminum paint that would not off-gas in a deep vacuum.[388] (NASA C–1961–58579)

Image 183: Twenty-seven-foot-diameter bulkhead at the north end of SPC No. 1. It contained a swinging 15-foot-diameter door in the middle that permitted the movement of large articles into and out of the chamber. (NASA C–1962–61466)

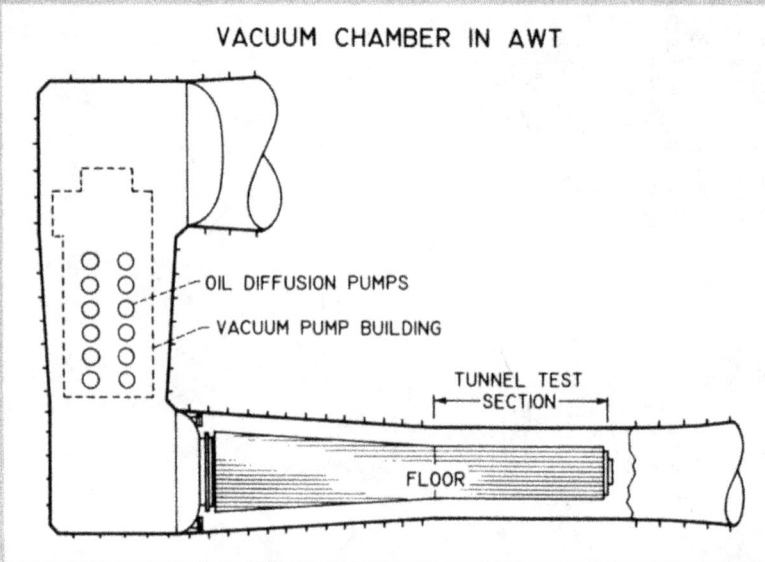

VACUUM CHAMBER IN AWT

OIL DIFFUSION PUMPS

VACUUM PUMP BUILDING

TUNNEL TEST
SECTION

FLOOR

Image 184: Diagram of SPC No. 1 showing the location of the Vacuum Pump House below the chamber. However practical it was to end the AWT's days as a tunnel, some, like NASA Lewis retiree Bob Walker felt differently, "When they welded it all and made a vacuum chamber out of it, they kind of destroyed the old feeling for it."[389] (NASA CS–1961–22859)

Image 185: Interior view of Vacuum Pump House being constructed below SPC No. 1. The portals for the 10 pending 32-inch-diameter oil-diffusion pumps can be seen in the ceiling of the pump house, which was directly underneath the chamber. (NASA C–1961–58582)

Image 186: SPC No. 1 in its initial configuration in the east leg of the tunnel. The 100-foot-long chamber was 31 feet in diameter at the southeast end, 27 feet in diameter at the northeast end, and had an internal volume of 70,000 cubic feet.[390] The diffusion pumps could reduce the atmosphere to a pressure level of 10^{-5} millimeters of mercury or roughly that found at an altitude of 100 miles. A removable dome would be added in the ceiling to lower a second-stage rocket into the tank. (NASA C–1962–61467)

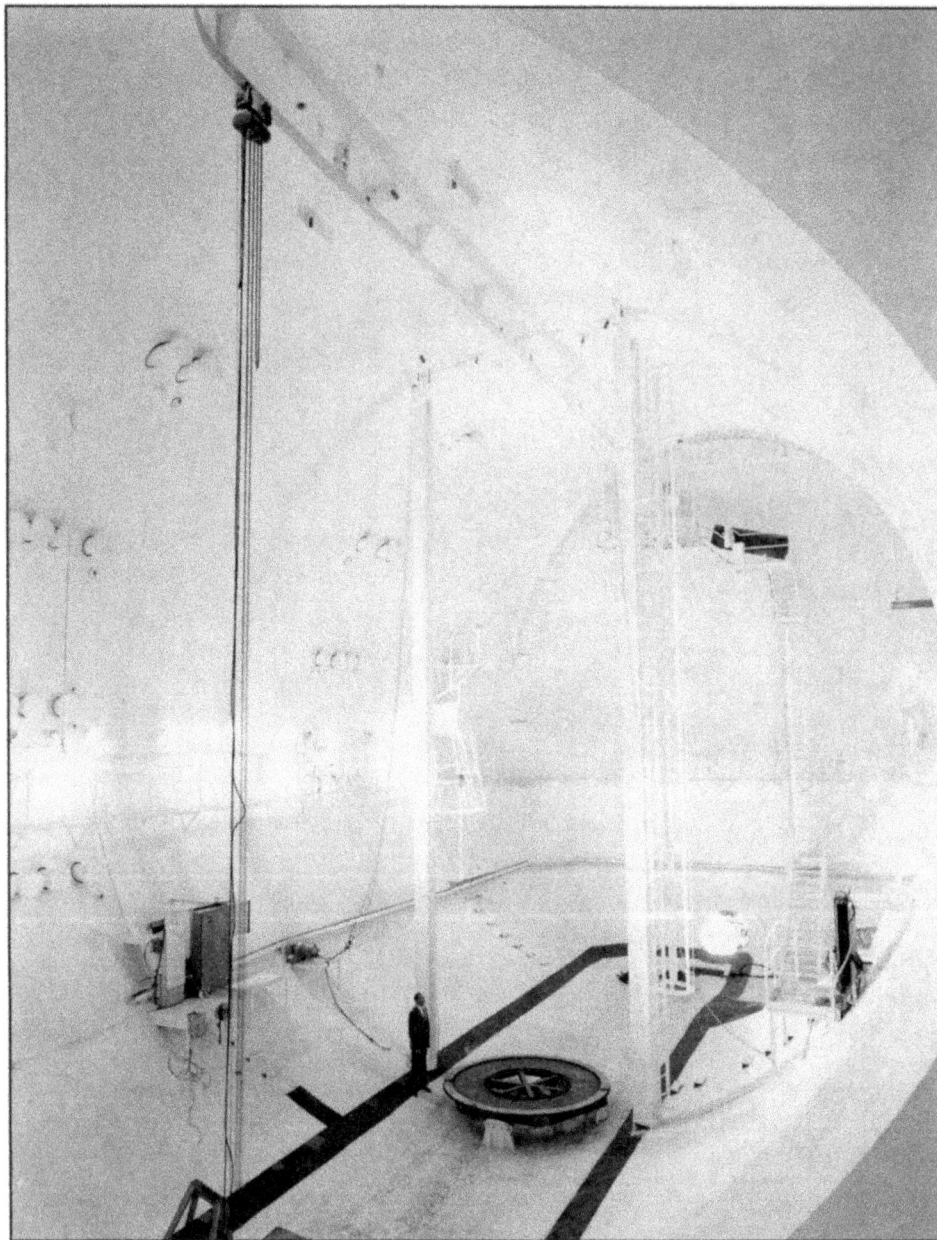

Image 187: SPC No. 2 in the western end and throat section of the tunnel. The section visible in this photograph was 51 feet in diameter and 121 feet long. The chamber also included the tunnel's throat section, which narrowed from 51 feet to 20 feet in diameter, and the back leg of the tunnel. The chamber used the existing tunnel exhaust systems to simulate altitudes of 100,000 feet. (NASA C-1965-00703)

Pulling Centaur from the Wreckage

On 8 May 1962, as construction of the SPC was nearing completion, the first Atlas/Centaur rocket lifted off from Launch Complex 36A at Cape Canaveral. The rocket rose into the clear sky and turned along its intended path. Then, as it neared the darkness of space at 49 seconds and 50 miles into the flight, the Centaur burst into a ball of flames.[391] The fireball streaked forward toward its destination for several seconds before falling into the ocean. In Cleveland, Lewis photographers were preparing to conduct a photographic survey of the center's new space tank. They were unaware of the effect that the Atlas/Centaur explosion would have on the SPC and the entire center.

Centaur was a 30,000-pound-thrust rocket designed in 1958 by General Dynamics as a second-stage for the U.S. Army's Advanced Research Projects Agency's Atlas intercontinental ballistic missile.[392] The Centaur Program was transferred in November 1959 to the Army Ballistic Missile Agency in Huntsville, Alabama. That agency, led by Werner von Braun, would soon be incorporated into NASA as the Marshall Space Flight Center.[393]

Centaur was a high-powered, sophisticated, and temperamental space vehicle. It used controversial, high-energy, highly explosive liquid hydrogen as its propellant. Centaur was the first cryogenic stage ever attempted and had unique balloonlike fuel tanks with a thin shared bulkhead between the oxidizer and propellant. It also had a two-burn capability, which meant it behaved both as a launch vehicle and as a spacecraft. Robert Gray, who directed many of the Cape Canaveral launches, explained to Virginia Dawson in 1999, "[The Centaur] had to be tweaked all the time; and it was not tolerant to any kind of failures, or something not being just right. It had to be right—period, or else it wasn't going to work."[394]

Yet Centaur's potential was great. "You could get more energy, more speed out of that Centaur than any rocket ever built up to that time. And perhaps even now," explained former Centaur launch analyst Joe Nieberding. "The Centaur enabled missions that would not have been flown; they were not doable. You could put a couple thousand pounds to Mars, whereas without the Centaur, you might only be able to put a couple hundred pounds."[395]

Initially, the Atlas/Centaur had a one-mission objective: *Surveyor*. NASA was relying on the new rocket to send a series of the *Surveyor* spacecraft to the Moon to photograph and sample the lunar surface as a prelude to the *Apollo* landings. Following the Centaur failure, there was debate within both Congress and NASA regarding Centaur's future. *Surveyor* was initially

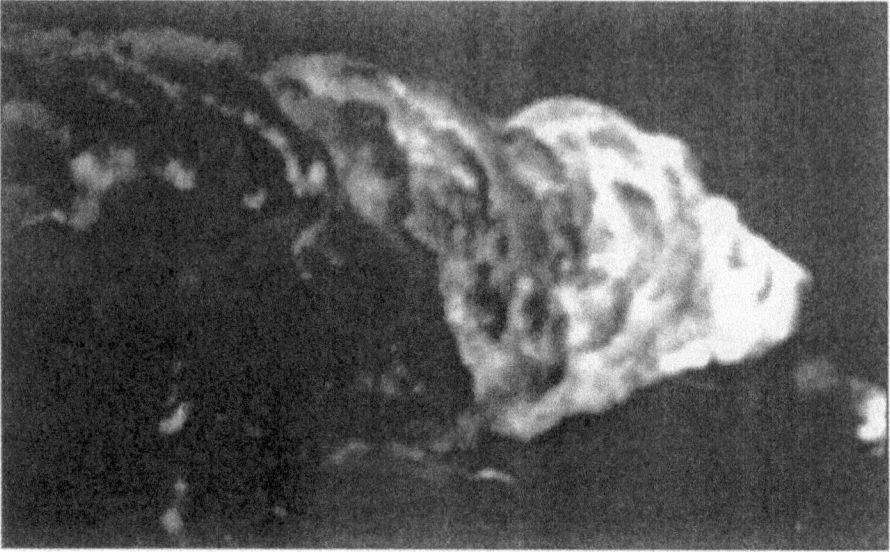

Image 188: The initial flight of a Centaur second-stage rocket with an Atlas booster was referred to as "F-1." It failed 49 seconds after liftoff on 8 May 1962. The explosion resulted from damage to one of the four panels that insulated the Centaur's cryogenic liquid-hydrogen propellant tank. The liquid hydrogen, which boils at −426°F, quickly expanded and ruptured the tank. The resulting explosion destroyed both the Centaur and the Atlas. (NASA MPD–1609)

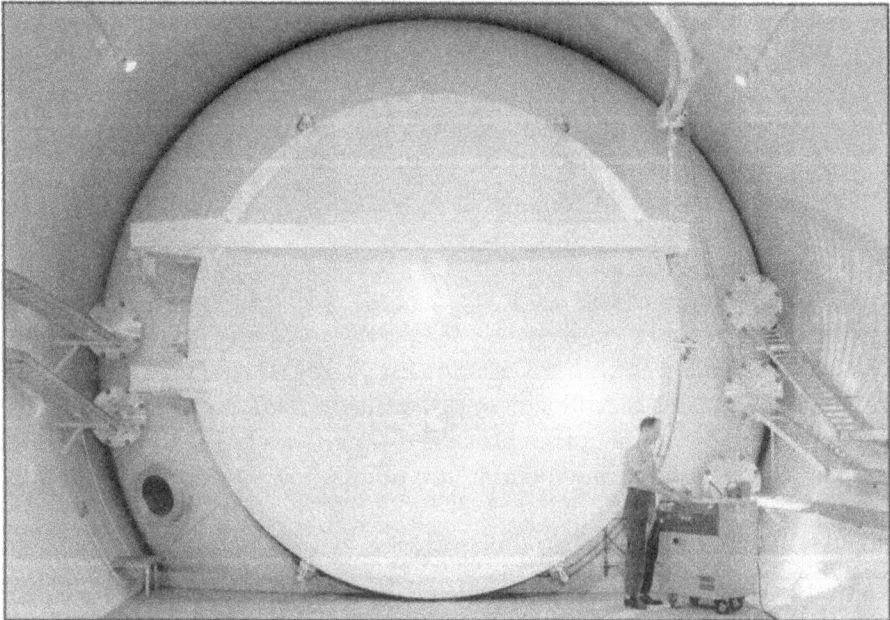

Image 189: Publicity photograph of a bulkhead in the new SPC taken two days after the Centaur F-1 explosion. Work would soon begin readying the unused facility for Centaur testing. (NASA C–1962–60341)

designed to include a Surveyor orbiter component, which would survey the Moon from a lunar orbit. After the Centaur failure, though, the orbiter was cancelled. It would be reconfigured as the Lunar Orbiter and launched on an Atlas/Agena.[396]

Hearings held by the House Subcommittee on Space Sciences following the F-1 failure found that General Dynamics had been responsible for the Centaur's design flaws, but that the "less than adequate" technical oversight by NASA Marshall had caused these flaws to go unnoticed prior to launch. Management at Caltech's Jet Propulsion Laboratory in Pasadena, California, where *Surveyor* was being developed, pressed hard for the substitution of the Atlas/Centaur with a Saturn C-1/Agena D rocket, as previously suggested by von Braun.[397] Because of its importance to the incipient Apollo Program,

Image 190: Centaur rockets in the General Dynamics assembly plant. General Dynamics had devised a new type of spacecraft that initially possessed a number of difficulties. It was the first cryogenic stage ever attempted, it would restart its engines in space, its tanks were balloons that had to remain pressurized, and there was a bulkhead the thickness of only a couple layers of aluminum foil between the oxidizer and propellant. (NASA C-1962-62408)

Centaur was not cancelled. It was, however, strongly recommended that the program undergo an intense reevaluation.[398] Ed Cortright, the Assistant Director of Lunar and Planetary Programs at NASA Headquarters, began taking steps to give Centaur a new start.

Nonetheless, the Jet Propulsion Laboratory, which was creating the *Surveyor* spacecraft that would be carried to the Moon, was irritated that the payload and its transportation, Centaur, were being developed simultaneously. The *Apollo* schedule was too tight for an uncertain rocket. The Jet Propulsion Laboratory management suggested that Centaur be cancelled even before the May 1962 failure of its first flight.[399]

In October 1962 the project was transferred to NASA Lewis. Over the next few years the entire Centaur Program was reevaluated at Lewis. The testing took place in both existing and specially constructed facilities. The Dynamics Research Stand was built at Plum Brook Station to test the structural fitness of the Atlas/Centaur. The Propulsion Systems Laboratory was used repeatedly to test Centaur's RL-10 engines and to study prelaunch chilldown of liquid helium. Finally, the newly created SPC was used for a wide variety of Centaur studies, including long-duration soaks in the new space tank and shroud separation tests.[400]

Accommodating Centaur

On 12 September 1962 the AWT was officially renamed the "Space Power Chambers" (SPC).[401] On 8 October, the actual transfer of the Centaur Program to Cleveland began. Almost at once, NASA Lewis management decided to modify the new SPC No. 1 to accommodate a full-size Centaur. Howard Wine, explained, "We didn't have another facility anywhere at Lewis that was capable of standing [the Centaur] up on its end, so that was logical. Otherwise they would have had either turned down the project, which was a real nice project to have because of high visibility, or they would have had to build another facility."[402] The SPC would be directly involved with 10 Centaur missions, and its tests would influence nearly every subsequent mission.

The decision to reconfigure SPC No. 1 for Centaur resulted in almost another year of construction on the unused space tank. The most time-consuming aspect was the addition of a 22.5-foot-diameter extension with a removable dome lid so that the Centaur could be set up vertically.[403]

Image 191: An extension and dome are added to SPC No. 1 to accommodate a full-scale Centaur rocket. A crane lifts the dome's lid into place as a welder works below shoring up the supports. This photograph was taken in August 1963, almost a year after the facility was supposedly completed. (7 Aug. 1963) (NASA C-1963-65692)

The Pittsburgh-Des Moines Steel Company, the same company that performed the initial AWT construction, was contracted in March 1963. Delays in the transportation of the steel pushed the construction into late June. Efforts, such as 60-hour work weeks and shipping steel on trucks rather than railroad, were made to expedite the work, but NASA officials were furious at the delays.[404] Pittsburgh-Des Moines argued that NASA's alterations to the design would have resulted in delays even if the steel had been delivered on time. The new dome was finally completed in August 1963.[405] After over two years of work, longer than the construction of the tunnel in the 1940s, the SPC was finally ready.

Centaur Comes to Cleveland

As NASA's Director of Space Flight Operations, Abe Silverstein led a committee in 1958 to explore the requirements for upper-stage rockets. The group concluded that high-energy propellants were a prerequisite. Concurrently, the army had agreed to General Dynamics's unique design of a liquid-hydrogen second-stage rocket called Centaur.[406] The following year, Silverstein established another team to select the upper stages for the new Saturn booster being developed at NASA Marshall. This group, known thereafter as the "Silverstein Committee," concluded that only upper stages using high-energy propellants would be practical. After a week of discussions, during which it is said that Silverstein reminded the group that the Project Bee liquid-hydrogen aircraft had been flown in Cleveland two years before, the reluctant von Braun agreed to the recommendations for the Saturn upper stages.[407]

Von Braun remained uncomfortable with the concept, however, and pushed to have the Centaur Program cancelled. He and his team of rocket experts in Huntsville had never felt at ease with Centaur's use of liquid hydrogen or its unconventional design. Andy Stofan, former Lewis Centaur manager, explained, "Literally it's pumped up like a balloon. If you let the air out, the tank collapses exactly like a balloon. Marshall built everything with I beams."[408]

Image 192. Silverstein and Werner von Braun on the 1958 Space Technology Committee. It has been assumed that the disdain between the two space titans stemmed from their ethnic backgrounds, but Silverstein claimed that any friction was the result of different management and budgetary styles, not personal animosity. (NASA C-1958-47955)

Ed Cortright obtained an informal agreement from Silverstein, who had become Director at Lewis, to take over the project if it was moved from Marshall. There are conflicting accounts of Silverstein's initial reaction to Cortright's suggestion that the project be transferred to Cleveland. It was clear, however, that both Silverstein and NASA Lewis had been at the vanguard of liquid-hydrogen development in the early 1950s. General Dynamics was familiar with Silverstein from his upper-stage committee in 1959 and pressed him to assume control of the project.[409]

Discussions on the future of Centaur were held in Huntsville. Even though von Braun still pressed for the project to be cancelled, he agreed to Cortright's proposed transfer.[410] Bruce Lundin, who oversaw all Lewis development work at that time, recalled that NASA Administrator James Webb summoned both Silverstein and him to Washington in August 1962, just weeks before the official opening of the SPC facility, to formally offer Lewis the project. Lundin recounted, "It was really a dog...miserable [in] technical, contractual, financial, in every which way it was in terrible shape." Lundin claimed later that he and Silverstein were among the few that saw it as an opportunity.[411]

In Cleveland, Silverstein would personally ensure that the Centaur would succeed and not delay the *Apollo* schedule. He hand-selected Lewis veterans such as Dave Gabriel, Cary Nettles, and Ralph Schmiedlin to run the project, but Silverstein would be heavily involved. Larry Ross explains, "[Silverstein] brought the program here because he felt it had to be done. And he felt he could do it... I have to infer this, but I think there was a matter of pride involved here. In as much as he said we could do it; he was going to make darn sure we did do it and we did it right."[412]

Image 193. Left to right: Abe Silverstein and Administrator James Webb. Silverstein left NASA Headquarters in 1961 because he was unhappy with Webb's restructuring of the space program. He assumed the position of Director at Lewis which had been vacant after Ray Sharp retired in January 1961. Silverstein recalled his return to Cleveland, "I was extremely happy to do that. In fact, I never wanted to leave."[413] (NASA C-1962-58736)

Endnotes for Chapter 8

368. Eugene Manganiello to I.H. Abbott and E.W. Conlon, 20 January 1960, NASA Glenn History Collection, Cleveland, OH.

369. Virginia P. Dawson, *Engines and Innovation: Lewis Laboratory and American Propulsion Technology,* (Washington, DC: NASA SP–4306, 1991), Appendix E.

370. Manganiello to Abbott and Conlon.

371. Interview with Larry Ross, conducted by Bob Arrighi, 1 March 2007, NASA Glenn History Collection, Cleveland, OH.

372. S. Choumoff, *Report on Certain Aspects of Space Simulation* (Washington, DC: NASA TT F–9574, 1965).

373. Lloyd Swenson, James Grimwood, and Charles Alexander, *This New Ocean: A History of Project Mercury* (Washington, DC: NASA SP–4201, 1966), chap. 9–2.

374. Interview with Abe Silverstein, conducted by John Sloop, 29 May 1974, NASA Glenn History Collection, Oral History Collection, Cleveland, OH.

375. John C. Sanders, "Aeronautical and Space Research Goals," 16 July 1959, NASA Glenn History Collection, Cleveland, OH.

376. David Namkoong, et al., *Photographic Study of Condensing Mercury Flow in 0 and 1-G Environments* (Washington, DC: NASA TN–D–4023, 1967), p. 11.

377. Sylvia Doughty Fries, *NASA Engineers and the Age of Apollo* (Washington, DC: NASA SP–4104, 1992), chap. 3.

378. R.T. Hollingsworth, *A Survey of Large Space Chambers* (Washington, DC: NASA TN D–1673, 1963), p. 12.

379. Schmiedlin, Ralph, et al., *Flight Simulation Tests of a Centaur Vehicle in a Space Chamber* (Washington, DC: NASA TM–X–1929, 1970), pp. 35 and 39.

380. Interview with Harold Friedman, conducted by Bob Arrighi, 2 November 2005, NASA Glenn History Collection, Oral History Collection, Cleveland, OH.

381. Interview with Bill Harrison, conducted by Bob Arrighi, 14 October 2005, NASA Glenn History Collection, Cleveland, OH.

382. Povolny, *Space Simulation,* p. 1.

383. Interview with Howard Wine, conducted by Bob Arrighi, 4 September 2005, NASA Glenn History Collection, Cleveland, OH.

384. Wine interview, conducted by Arrighi, 4 September 2005.

385. Wine interview, conducted by Arrighi, 4 September 2005.

386. Wine interview, conducted by Arrighi, 4 September 2005.

387. Jill Taylor, "The NASA Glenn Research Center: An Economic Impact Study Fiscal Year 2006," September 2007, NASA Glenn History Collection, Cleveland, OH.

388. John Povolny, *Space Simulation and Full-Scale Testing in a Converted Facility* (Washington, DC: NASA N66 35913, 1970), p. 2.

389. Interview with Bob Walker, conducted by Bob Arrighi, 2 August 2005, NASA Glenn History Collection, Cleveland, OH.

390. Demolition of Building #7: Altitude Wind Tunnel. General Arrangement Plan AWT Base Bid/Option 3, drawing ED602, NASA Glenn History Collection, Cleveland, OH.

391. Ed Kyle, "Atlas Centaur LV–3C Development History," *Space Launch Report* (28 May 2005), http://www.spacelaunchreport.com/ (accessed 2 September 2009).
392. "Report to the Committee on Science and Astronautics House of Representatives. Review of the Atlas Centaur Launch Vehicle Development Program, NASA," March 1963, NASA Glenn History Collection, Cleveland, OH.
393. Bruce Lundin, "Director Praises Centaur Program," *Lewis News* (20 October 1972).
394. Interview with Robert Gray, conducted by Virginia Dawson, 9 November 1999, NASA Glenn History Collection, Cleveland, OH.
395. Interview with Joseph Nieberding, conducted by Virginia Dawson and Mark Bowles, 15 April 1999, NASA Glenn History Collection, Cleveland, OH.
396. Brian Sparks to Homer E. Newell, 21 September 1962, NASA Glenn History Collection, Cleveland, OH.
397. Sparks to Newell.
398. "Report to the Committee on Science and Astronautics."
399. Virginia Dawson and Mark Bowles, *Taming Liquid Hydrogen: The Centaur Upper Stage Rocket, 1958–2002* (Washington, DC: NASA SP–4230, 2004), p. 55.
400. "1962–1972 The First Decade of Centaur," *Lewis News* (20 October 1972).
401. Abe Silverstein to staff, "Updating the Name of Structure Number 7 of the Lewis Research Center," 12 September 1962, NASA Glenn History Collection, Cleveland, OH.
402. Wine interview, conducted by Arrighi, 4 September 2005.
403. Povolny, *Space Simulation,* p. 1.
404. R.W. Bouman, "Contracting Officer's Decision, Contract NA53–3434," February 28, 1964, NASA Glenn History Collection, Cleveland, OH.
405. John Dickson, "Periodic Status Report, Contract NAS3–3434, Project Number C–3674 C–1," 29 August 1963, NASA Glenn History Collection, Cleveland, OH.
406. "1962–1972 The First Decade of Centaur."
407. Dawson and Bowles, *Taming Liquid Hydrogen,* p. 35.
408. Interview with Andrew Stofan, conducted by Tom Farmer, *This Way Up: Voices Climbing the Wind,* WVIZ Documentary, 1991, NASA Glenn History Collection, Cleveland, OH.
409. Interview with Grant Hansen, conducted by Virginia Dawson and Joe Nieberding, 6 June 2000, NASA Glenn History Collection, Cleveland, OH.
410. Interview with Edgar M. Cortright, conducted by Rich Dinkel, 20 August 1998, Johnson Space Center Oral History Project, Houston, TX.
411. Interview with Bruce Lundin, conducted by Tom Farmer, *This Way Up: Voices Climbing the Wind,* WVIZ Documentary, 1991, NASA Glenn History Collection, Cleveland, OH.
412. Interview with Larry Ross, conducted by Tom Farmer, *This Way Up: Voices Climbing the Wind,* WVIZ Documentary, 1991, NASA Glenn History Collection, Cleveland, OH.
413. Silverstein interview, conducted by Sloop, 29 May 1974.

Image 194: The first study in the new Space Power Chambers used a unique model that simulated the Atlas/Centaur's size and weight to test retrorockets. During this time a full-scale Centaur rocket was being readied for testing in the facility's vacuum chamber. (NASA C–1964–69428)

Chapter 9

Space Power Chambers Join the Fray | Early Centaur Period (1963–1964)

On 8 October 1962 the storm hit. "Communications were flying throughout the NASA, most of which I did not see until after," former Centaur Program Assistant Manager Cary Nettles recalled in 2002. "I found out quickly just what a big mess we had been handed." Abe Silverstein had assigned Nettles the daunting task of assembling the Centaur Project Office at the NASA Lewis Research Center. Engineers from the NASA Marshall Space Flight Center, annoyed that they had been ordered by NASA Headquarters to assist with the transition, arrived in Cleveland hoping to make their stay as short as possible. The boxes of paperwork that arrived from Huntsville revealed that General Dynamics had hundreds of updates for the Centaur design that had not yet been implemented. Management of the program was fragmented. The air force oversaw the Atlas booster, Marshall the Centaur's RL-10 engines and launch pad, and Lewis the Centaur stage. In addition, Lewis had never managed a developmental program of this size or with this amount of national visibility.[414]

The Space Power Chambers (SPC) would not be ready for the first two Lewis-run launches, but it would contribute significantly to three of the four ensuing Centaur missions. While waiting for the SPC to be modified for the Centaur testing, large test articles were delivered to the facility. Lewis personnel would spend the subsequent months studying the equipment to learn about both the Centaur rocket and spaceflight hardware in general. When the SPC was finally ready in September 1963, the staff wasted no time getting started.

Lewis Takes Control

During its first year with the Centaur Program, Lewis had undertaken an intensive redesign and testing effort. By the time that the SPC was completed in late August 1963, Lewis had already modified the Centaur in several significant ways. These included implementation of a comprehensive ground-testing program, use of a low thrust level during coast periods, a smaller oxidizer tank, ground-chilled liquid helium, and an improved electronics programmer. The most significant change was the temporary switch from a parking orbit to a direct-ascent trajectory until Centaur had proven itself. The direct ascent would pose fewer technical problems but would reduce the available launch opportunities by almost 75 percent.[415]

Image 195: Front desk: Ralph Schmiedlin, Head of the SPC Section of the Test Engineering Branch, at his desk in the Blockhouse. Center and back desks: Harold Groth and Ben Dastoli. Schmiedlin had transferred from NACA Langley in 1943.[416] His group was responsible for developing and running propellant dynamics and thermo-dynamic tests in a space environment for Centaur. (1963) (NASA MPD–538)

Just over a year after the Marshall team unloaded the boxes of Centaur change orders, drawings, and other documents on the Lewis tarmac, Centaur was ready for another launch attempt. The 27 November 1963 launch successfully placed a mock-up payload into orbit. Atlas/Centaur 2 (AC-2) became the first liquid-hydrogen rocket to reach space. Nettles recalled, "And 70 seconds after liftoff we discovered that Centaur didn't break up, and we had a chance to continue the project."[417]

Six months later when Centaur's follow-up single-burn launch suffered a hydraulic malfunction, the Lewis group was again under the gun. They were already looking to the late-1964 AC-4, which would be Centaur's most complex flight to date. It would include the first restarting of the Centaur's engines in space, the first use of the inertial guidance system, and the first *Surveyor* mass model as its payload.[418] Although the SPC was not ready in time for Lewis's first two Centaur attempts, the facility would be used at length in preparation for AC-4.

Extreme Diligence

Tests were developed by Centaur Program Manager Dave Gabriel and Cary Nettles, but Center Director Abe Silverstein personally watched over much of the early Centaur work. Larry Ross stated, "There's no doubt in my mind that [Silverstein] approved every test concept. He was in many ways the de facto Centaur Program Manager.... All the managers on Centaur would have to come in on Saturday morning and stand and deliver."[419] Nettles informed Virginia Dawson, "This management setup was a good one and was certainly largely responsible for the success of the Centaur Program. One of the interesting aspects of this arrangement was that Abe never did really give up the idea that I reported directly to him. I would receive direct calls from him asking about various aspects of the program."[420]

Silverstein's solution to technical problems was a rigorous testing program. He insisted that the Centaur's components, subsystems, systems, and entire vehicle be subjected to extensive ground testing. He felt that nothing should fly in space without proving itself under similar conditions on the ground. He later explained, "This 100-percent [reliability] requirement is of course almost required in space activity, and we get it not by anything except extreme diligence and the concept that every piece of equipment that is taken aboard, every component must be proven, environmentally checked so that it can live in the environment of space, the total environment of space, the vacuum of space, the temperatures of space, the pressures of space; that is, in the fields of other components; that the whole ensemble put together is tested environmentally so as to operate successfully."[421] The SPC was created for just this purpose.

Centaur's Test Engineering Branch was divided into two sections—Plum Brook Station's E Test Stand for full-scale structural dynamics studies and the SPC for tests in a simulated space environment. The SPC section was headed by NACA veteran Ralph Schmiedlin. For the AC-4 mission, three test programs were developed for the SPC: the Atlas/Centaur separation system, the behavior of electrical and guidance systems after long durations in a space environment, and the jettisoning of the nose cone for Centaur's Surveyor payload.

Image 196: Lewis personnel check out the Atlas/Centaur mass model inside SPC No. 2 on August 1963. The model, delivered the previous day on a small convoy of trailers, was promptly inserted into the chamber to begin setup for the separation system tests. Workers were continuing to assemble the dome on SPC No. 1 at the other end of the facility. (NASA C–1963–65614)

Whale-Bone

Engineers on loan from General Dynamics worked closely with the Lewis team throughout the late summer and fall of 1963 preparing for the initial Centaur tests in the SPC. Their first objective was readying the Atlas/Centaur mass model in SPC No. 2 for a sequence of separation tests. A shaped explosive cord was wound around the rocket's interstage adapter. Once the cord was exploded, eight small retrorockets pulled the Atlas away from the Centaur.[422] Although General Dynamics did not foresee any problems with the separation system, the Lewis group, which was much more hands-on than NASA Marshall had been, wanted to verify the flight dynamics with a full-scale model and test its subsystems. This study would result in a new jettison method that would significantly reduce the separation time and thus minimize the danger of collision between the two stages during separation.[423]

Image 197: Atlas/Centaur mass model suspended from the trolley system inside SPC No. 2. The dark section to the right is the fixed Centaur model. The corrugated section above the researchers is the interstage adapter that was jettisoned from the Centaur using linear-shaped charges. The long cagelike Atlas portion of the model begins just to the left of the adapter and continues out of view to the left. Eight retrorockets were used to push this section away from the Centaur.[424] (NASA C–1963–65907)

The "whale-bone" Atlas/Centaur mass model was suspended horizontally on a trolley system inside the chamber with a net hung at one end to catch the jettisoned Atlas model. The chamber atmosphere was reduced to a pressure altitude of 100,000 feet, and high-speed cameras were synchronized to the ignition of the retrorockets. After a number of test runs from late September to mid-November 1963, the Lewis researchers on the team, Henry Synor and Dick Heath, determined that the cord-shaped charge performed well but that the firing of the retrorockets was seriously inconsistent. In early 1964, the team conducted tests again using both standard and alternative versions of the rocket igniters. The new studies determined that the firing problems resulted from the igniter's unpredictability and the shortness of the burning period.[425]

Responding to Lewis's suggestions, the rocket manufacturer developed an improved igniter that was fired electronically. When tested in the SPC, the

Image 198: The Atlas portion of the model separates from the simulated Centaur. Following the initial findings in late 1963, Rocket Power, Inc., developed its own improved igniter. The system was again tested during the spring of 1964 in SPC No. 2. The extended-burn igniters used an initiator and fired electronically. Propellant granules used in the sustainer charge improved the heat input by providing hot gases to the grain. The researchers found that the newly designed igniter fired properly using any closure thickness.[426] (NASA C-1964-68534)

Image 199: Left to right: K. G. Smits, R. G. Sims, and J. E. Rogers of General Dynamics's Astrophysics Division prepare for an Atlas/Centaur retrorocket test in the SPC No. 2 control room. General Dynamics trained the Lewis staff on the Centaur for the initial testing.[427] (NASA C–1963–66620)

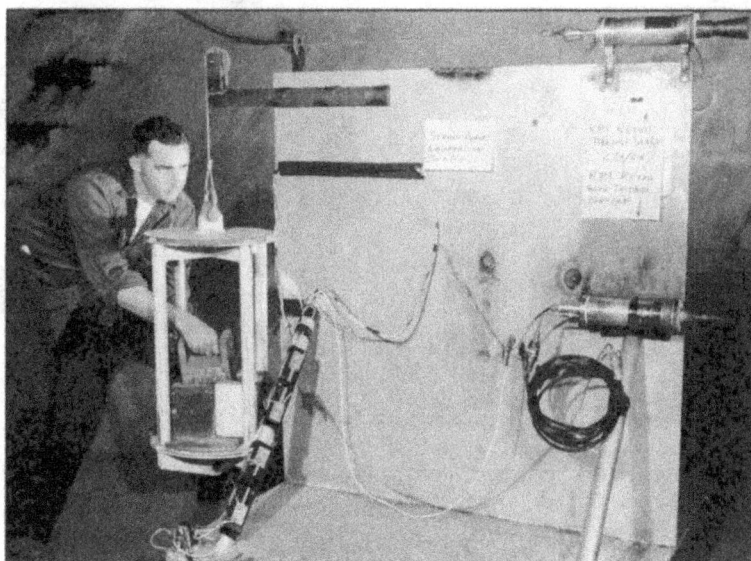

Image 200: The retrorockets were tested with their igniters packed with various forms of boron potassium nitrate. The igniters failed six times during their 67 firings. Researchers also used foam panels to record the flame patterns and found a wide range of disbursement among the various igniters. A new igniter, which directed its flame directly at the propellant and burned for a longer duration, was selected.[428] (NASA C–1964–70158)

Image 201: Center: Lewis researcher Henry Synor goes over the Atlas/Centaur test setup with General Dynamics technicians inside SPC No. 2. (NASA C–1963–65906)

Image 202: In late 1963 and early 1964, local radio station WHK featured NASA Lewis and Plum Brook Station during several episodes of its "Highlights in Education" series. The February broadcast dealt with the Centaur work performed at Lewis and in its SPC facility. In this photograph, Ralph Schmiedlin (center) explains the Atlas/Centaur retrorocket test to the reporters. They are standing on a platform in SPC No. 2 with a Centaur mass model behind them. (NASA C–1963–67456)

redesigned retrorockets performed well, even when one retrorocket malfunctioned.[429] In addition, Dick Heath and Henry Synor reconfigured the propellant load to increase the rockets' impulse and thus accelerate the separation. Larry Ross recalled, "Those tests really did change the fundamental design of the separation system."[430]

Sick Patient Arrives

On 27 September 1963, just days before the first Atlas/Centaur separation test was run in SPC No. 2, a C-130 aircraft arrived from Cape Canaveral on the NASA tarmac and unloaded a Centaur 6A rocket on a stretcher. Cary Nettles recalled, "We did, indeed, inherit a pretty sick patient at the time. Technical flaws and mistakes were evident in many places, and, actually, it was something of a mess. It had many unclear goals and program objectives...our contactor was confused at best and demoralized at worst."[431]

Image 203: The Centaur 6A rocket is delivered to NASA Lewis for testing in SPC No. 1. Because of its fragile structure, the Centaur had to be transported on a stretcher to make sure that it did not collapse on itself. It was flown from Cape Canaveral on a C-130 transport aircraft. ("Centaur engine arriving at Lewis aboard C-130," 27 September 1963, NASA MPD-498) (NASA Glenn film collection)

The rocket, which was to be used for the space environment tests, was brought into the SPC high bay where it was stood up vertically. Immediately Lewis engineers and technicians began working with it. Since the rocket was a leftover single-burn Centaur, the electronics and operating systems had to be updated to the two-burn AC-4 configuration.

Again the General Dynamics advisors were involved with not only preparing the rocket for testing but also teaching the NASA personnel how to work with spacecraft technology. Months were spent reharnessing the Centaur's electronics, learning about the systems, and being taught how to handle flight hardware. Ross explained, "Our people came out of a history of putting a scale model in the test section of a wind tunnel and playing with it; and when it didn't yield the right data, then messing around with it until it did. That's not the way you handle flight hardware. So we had a learning experience to go through."[432]

Image 204: A team from General Dynamics was brought in to train Lewis personnel, many of whom had little space experience, on the complex Centaur rocket. This Centaur was an early 6A model that was originally slated as a follow-up to the AC-2 flight. A rehabilitation of the pad delayed the launch, however, and in the interim an updated Centaur was used instead for the AC-3 launch.[433] (NASA C–1963–66499)

During this period, SPC No. 1 was outfitted with several additional components that would enhance the space simulation for Centaur. The rocket was surrounded by a large "cold wall" that was used to simulate the cryogenic temperatures of space. Solar radiation was reproduced using two-hundred fifty-nine 500-watt lamps arranged in six arrays.[434] Although the two RL-10 engines would not be fired during these tests, a hydraulic system rotated engines as they would when steering the rocket during a mission. Microgravity, motion, and flight vibrations were the only aspects of spaceflight that were not recreated.[435] Ross explained, "It was Silverstein doctrine that if it could be tested, you could simulate it. It may not be

Image 205: Lewis technicians study the Centaur's RL-10 engines in the SPC shop. Centaur was the center's highest profile program, and the SPC was an important part of the program. Abe Silverstein felt, "If the space program did nothing else than to show that if you want to develop a good product for a given use, whether it be on the ground or in space, the thing you do is to design it and develop it for the environment in which it is going to live."[436] (NASA C–1964–71100)

perfect, you certainly have one G, but you could do it in a vacuum. You can get a lot of feeling for how the system's going to behave in flight."[437]

Centaur was a fragile spacecraft with thin inflated tanks that had to be pressurized to retain their proper form. If the pressure difference between the upper section of the vehicle and the lower portion changed, the vehicle would likely collapse. A pneumatic system was installed in the chamber to maintain this pressure differential, and the vehicle pressurization was monitored around the clock seven days a week.

Howard Wine, a technician in the SPC at the time, remembered, "So it was a whole new experience for us. We never had our hands on that kind of sophistication...and so everybody is walking around on egg shells...because [the Centaur] was so thin, so delicate."[438]

By early spring 1964, the extensive setup of both the spacecraft and the chamber was finally completed. On 19 March the Centaur was rolled out from the shop, hoisted high into the air by a crane, and lowered into the waiting space tank. Nearly three years after construction started, the center's new vacuum chamber, SPC No. 1, was finally ready to be used.

Learning New Behaviors

NASA Lewis had almost doubled its staff in 1962 and 1963, and by 1964 Lewis had become NASA's second largest center.[439] Most of the new recruits were young and just out of university. The center had already been expanding its research in the nuclear, chemical, and ion propulsion fields, as well as in power conversion. Three large space-development programs were transferred to Cleveland in the early 1960s—the M-1 engine and the Agena and Centaur rockets. The impact of the programs on the center was enormous and affected a large swath of Lewis's resources.[440]

A two-story cinderblock office building was built inside the center's aircraft hangar to house the staff for the new development programs during their crucial initial years. This "Blockhouse," just a couple of minutes walk from the SPC, would be the hub for Lewis space engineers and managers during Centaur's formative years. By mid-1965, almost 1,000 employees had been transferred to the new Development Engineering Building and its annex outside of the main campus area.[441]

During the transition of the Centaur Program from Huntsville to Cleveland, NASA Marshall personnel were prohibited from transferring out of the Centaur Project and were required to participate in a temporary detail to Lewis.[442] Vernon Weyers, a former Centaur launch vehicle mission analyst, remembers the atmosphere in the early days of Centaur. "Every Monday morning, an airplane would pull up on the tarmac and all these Marshall engineers would file off, and they would spend the week, and on Friday some time, the plane would go back to Huntsville for the weekend."[443]

The Marshall engineers were eager to permanently return to Huntsville. Lewis was soon left on its own. The pressure on the Lewis office was intense during Centaur's first years. Lewis had to not only create new test facilities and remedy the Centaur's technical flaws but also learn how to work with space hardware. A team of 25 to 30 engineers from General Dynamics, the designer and manufacturer of Centaur, were transferred from San Diego to Cleveland for two years to teach the Lewis group about Centaur. Larry Ross explained, "There was a fair degree of the NASA people having to learn behaviors. We were inclined, or the NASA people were inclined to do tests in a very seat-of-the-pants way. That's not the way you handle [space] flight hardware. Flight hardware is careful controls, inspectors watch everything you do, carefully maintain configuration management."[444]

Lewis also had to learn to manage a large development program that included the military and contractors. NASA would have to adapt to the different cultures of these groups. Seymour Himmel explained that the air force liked to give contractors the work and almost all responsibility for developing an aircraft or rocket. "[Lewis] was coming from basically a hands-on research kind of an organization; we couldn't do anything we couldn't put our hands on."[445] Meanwhile, General Dynamics, which had been almost entirely responsible for the program, found Lewis's extra involvement frustrating at first, particularly its extensive test and qualification program. Soon, however, both sides would come to appreciate the abilities of the other.

Image 206: As Lewis shifted from aeronautics research to spacecraft development work in the early 1960s, the center's engineers and mechanics had to alter the way that they approached tests. Aircraft engines could often be tinkered with to improve performance, but space hardware required precision and adherence to specifications. Here a Surveyor nose cone is prepared in the high bay for separation tests in SPC No. 1. (NASA C–1964–70592)

Endnotes for Chapter 9

414. Cary Nettles to Virginia Dawson, "Centaur Notes," 8 May 2002, NASA Glenn History Collection, Cleveland, OH.
415. *Surveyor Project Review: August 27–28, 1963* (Pasadena, CA: JPL CR–95760, 1966), pp. 9–10.
416. "Many End Careers at Year's End" *Lewis News* (10 January 1979), p. 3.
417. Cary Nettles, "Talk at NASA Group Achievement Award Ceremony at NASA Lewis Research Center," 1 December 1966, Cleveland, OH.
418. "3 Firsts Set for AC–4 Test Flight," *Lewis News* (11 December 1964), p. 1.
419. Interview with Larry Ross, conducted by Bob Arrighi, 1 March 2007, NASA Glenn History Collection, Cleveland, OH.
420. Nettles to Dawson, "Centaur Notes."
421. Interview with Abe Silverstein, conducted by John Sloop, 29 May 1974.
422. "Space Vehicles," Inspection of the Lewis Research Center, 1966, NASA Glenn History Collection, Cleveland, OH.
423. "1962–1972 The First Decade of Centaur," *Lewis News* (20 October 1972).
424. Richard Heath, et al., *Investigation of Atlas Solid Fuel Retarding Rocket During Atlas-Centaur Separation Tests* (Washington, DC: NASA TM–X–1119, 1965), p. 5.
425. Richard Heath, et al., *Study of Separation Dynamics and Systems Related to Staging of Atlas-Centaur* (Washington, DC: NASA TM–X–52292, 1967), pp. 2 and 5.
426. Richard Heath, Henry Synor, and Ralph Schmiedlin, *Test of Improved Igniter for First-Stage Separation Rockets for the Atlas-Centaur Launch Vehicle* (Washington, DC: NASA TM–X–1676, 1968).
427. Ross interview, conducted by Arrighi, 1 March 2007.
428. Heath et al., *Investigation of Atlas*, p. 4.
429. Heath et al., *Test of Improved Igniter*.
430. Ross interview, conducted by Arrighi, 1 March 2007.
431. Nettles, "Talk at NASA Group Achievement Award Ceremony."
432. Ross interview, conducted by Arrighi, 1 March 2007.
433. Ralph Schmiedlin, et al., *Flight Simulation Tests of a Centaur Vehicle in a Space Chamber* (Washington, DC: NASA TM–X–1929, 1970), pp. 12 and 39.
434. Schmiedlin et al., *Flight Simulation Tests*, pp. 12 and 35.
435. John Povolny, *Space Simulation and Full-Scale Testing in a Converted Facility* (Washington, DC: NASA N66 35913, 1970), p. 49.
436. Silverstein interview, conducted by Sloop, 29 May 1974.
437. Ross interview, conducted by Arrighi, 1 March 2007.
438. Interview with Howard Wine, conducted by Bob Arrighi, 4 September 2005, NASA Glenn History Collection, Cleveland, OH.

439. "Lewis Research Center," NASA Lewis press release, 17 January 1964, Public Information Office, Lewis Research Center.
440. Nettles to Dawson, "Centaur Notes,"
441. "Occupancy of DEB Annex is Slated for Near Future," *Lewis News* (8 January 1965), p. 1.
442. Robert C. Seamons to Abe Silverstein, "Atlas Centaur Success. The Perfection of the Centaur AC–10," 22 October 1962, NASA Glenn History Collection, Cleveland, OH.
443. Interview with Vernon Weyers, conducted by Mark Bowles, 8 April 2000, NASA Glenn History Collection, Oral History Collection, Cleveland, OH.
444. Ross interview, conducted by Arrighi, 1 March 2007.
445. Interview with Seymour Himmel, conducted by Virginia Dawson, 1 March 2000, NASA Glenn History Collection, Cleveland, OH.

Image 207: A full-scale Centaur second-stage rocket is lowered into the Space Power Chambers' vacuum tank. The facility was used extensively for a number of tests to prepare Centaur for its critical missions to send Surveyor spacecraft to the Moon. (NASA C–1964–68846)

Chapter 10

Ad Astra Per Aspera | The Centaur/Surveyor Missions (1964–1967)

The Cleveland area public was hungry for NASA news in the early 1960s. Local media outlets created in-depth special reports on NASA Lewis Research Center's contributions to the space program, the center's speakers bureau was in high-demand, and for the first time, large space expositions were held. In April 1961, 12,500 Clevelanders attended a display of satellites at the Case Institute of Technology (now called Case Western Reserve University), and an August 1962 open house attracted 17,000 visitors to Lewis. Most impressive of all was the enormous "Space Science Fair" at Cleveland's Convention Center. The event, cosponsored by NASA and The *Cleveland Plain Dealer,* drew an estimated 300,000 people in the fall of 1962.[446]

Gordon Cooper flew the final Mercury mission on 15 May 1963; it would be nearly two years before humans would enter space again. In this lull between Mercury and Gemini, the public's attention turned to a series of uncrewed missions that would explore the Moon in preparation for the eventual Apollo flights. The two primary phases were the Ranger and Surveyor programs. The Centaur second-stage rocket would be used to launch Surveyor's five test flights and seven actual missions. The *Surveyor* spacecraft would perform the first controlled landings on another planet. Centaur was one of the most complex and temperamental rockets ever conceived. It had been rife with problems when the program was adopted by Lewis in 1963. Centaur was the center's highest profile program during the Agency's highest profile years.

Image 208: From 1962 to 1964, NASA Lewis had a better relationship with the press and community than at any other period.[447] *Lewis was the largest of over 300 participants in the Cleveland Press Parade of Progress Exposition held 28 August through 7 September 1964 at Cleveland's Public Hall. A model of a Centaur rocket with its Surveyor payload is seen to the left in this photograph. (NASA C–1964–71681)*

Pathway to the Moon

The Surveyor Program was begun at the Jet Propulsion Laboratory in May 1960. Initially it had no association with the Apollo Program. During the initial planning stages, NASA intended to launch the Surveyors with the Atlas/Agena vehicles being used for the lighter Ranger missions. By July 1962 NASA's engineers determined that the Agena would not be powerful enough to carry the *Surveyor*. The Atlas/Centaur would be used instead.[448] Centaur, whose first launch had exploded in the atmosphere just two months before, was now a vital component of the $469 million Surveyor Program and thus critical to the design and strategy for the Apollo missions.[449] In October the Centaur Program was given a second chance with its transfer to Lewis.

The Centaur/Surveyor Program included eight developmental flights: three Atlas/Centaur tests, one mass model payload, and four *Surveyor* mock-ups used to verify that the vehicle could successfully put a spacecraft on a trajectory toward lunar orbit. It was essential to approach the Moon at the right speed so that the retrorockets could set the vehicle down gently.[450] It was also important to time the landing so that the *Surveyor* would have the maximum amount of sunlight for its solar panels.[451] Centaur, with its ability to restart its engines in orbit, would provide some flexibility in scheduling the landings. Silverstein had decided to forgo the restarting of the engines during the previous two launches so that Lewis engineers could concentrate on Centaur's other problems. The upcoming launch of Atlas/Centaur 4 (AC-4) in late 1964 would not only carry the first dummy *Surveyor* payload but also be the first two-burn mission.

Centaur's Simulated Flight

Centaur included a number of systems and subsystems whose behavior in space was critical to the success of the mission. The electronics and control systems were at the forward section of the rocket, just below the payload, and the mechanical and propulsion systems were at the base.[452]

Image 209: The Surveyor landing vehicle was basically a large tripod with 4.3-meter legs. Different equipment packages were used for the various missions. The early missions contained only cameras, whereas the later spacecraft also included excavating equipment. The Surveyor III spacecraft shown here landed on the lunar surface on 20 April 1967. In this photograph, which was taken in November 1969 by the Apollo 12 crew, a 1-meter mast with a solar cell panel rises above the base.[453] (NASA AS12–48–7121)

In preparation for the two-burn flights, Cary Nettles, who managed the Centaur's electronics and ground support equipment, wanted to run a thermal vacuum test to verify these systems, particularly the autopilot and guidance arrangements.[454]

Electronic malfunctions were the most likely cause of failures in space. Studying the electrical and electronics systems during long soaks inside the SPC No. 1 space tank would help the Lewis team calibrate the systems and facilitate the monitoring of the spacecraft during an actual flight. Silverstein, who remained very involved with the program, felt that it was important to use a fully operational Centaur for the tests so that the interaction of the different electronics could also be studied.[455]

A Centaur with an AC-4 configuration was mounted vertically in the chamber and subjected to a series of 60-minute low-Earth-orbit simulations. For each test, the systems were first given a dry run in a normal sea-level atmosphere. Then, the chamber was sealed and the vacuum pulled down, liquid nitrogen was used to chill the cold wall and partially fill the propellant tanks, the Centaur was brought to launch temperature, the electrical umbilicals were turned on, and the test commenced.[456]

The first 3 minutes of the test replicated the Atlas booster phase. The Centaur electronic systems were activated during a simulated separation. The electronics systems were studied throughout the remainder of the virtual mission: prestarting the RL-10 engines, simulating the engine ignition and cutoff, coasting for approximately 25 minutes, then simulating a second engine ignition and cutoff, payload separation, Centaur course reversal, and finally the shutdown of all Centaur systems.[457]

Photo Essay 6:
How Centaur Was Studied in Space Power Chamber No. 1

Image 210: A 100-foot crane is used to lower a Centaur 6A into SPC No. 1 and onto a waiting stand. The dome and removable lid were added to the facility specifically to accommodate the 28.5-foot-long, 10-foot-diameter Centaur vehicle. The rocket included autopilot, guidance, main propulsion, hydraulic, hydrogen peroxide supply, boost-pump attitude control, telemetry, tracking, range safety, and pneumatic systems. The electronics and control systems were at the forward section of the rocket, and the mechanical and propulsion systems were near the rear. (NASA C–1964–68844)

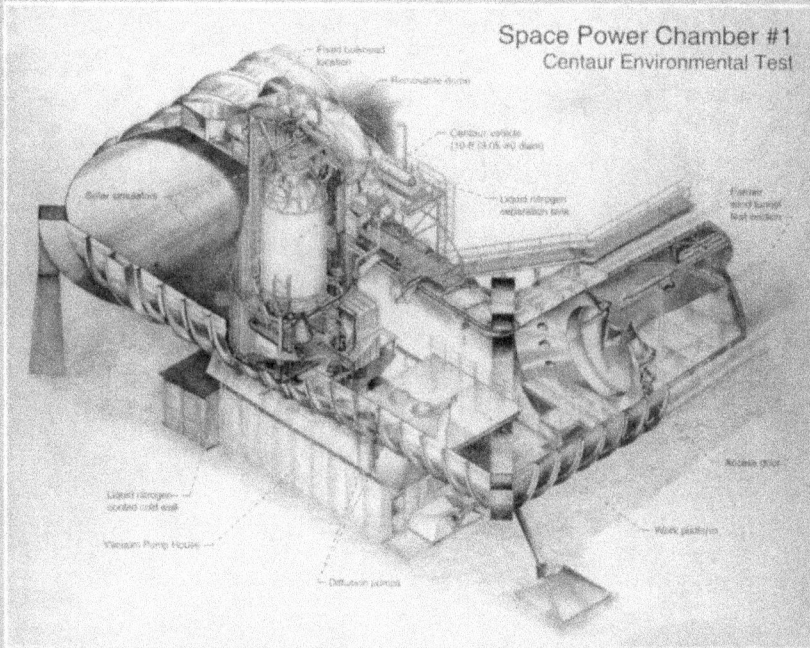

Image 211: Cutaway drawing of SPC No. 1 with the Centaur test setup. (NASA P–1047)

Image 212: The 22.5-foot-diameter dome being placed onto SPC No. 1. Instrument portals can be seen at the fixed-base of the dome. The top of the ribbed cold wall is visible inside the lid. (NASA C–1964–67913)

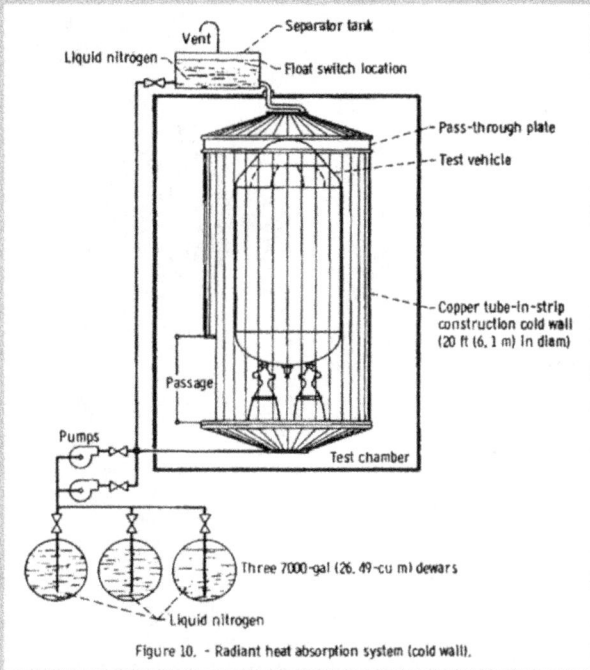

Figure 10. - Radiant heat absorption system (cold wall).

Image 213: A large copper liquid-nitrogen-cooled baffle, 20 feet in diameter and 42 feet high, was erected around the entire Centaur setup to simulate the temperatures of outer space. The liquid nitrogen, which was stored in three 7,000-gallon tanks, flowed through a separation tank that automatically kept the cold wall filled. (NASA TM X–1929, fig. 10)

Image 214: A radiant heater system was designed specifically for the tests. Six sectors of 500-watt tungsten-iodine lamps were arranged around the rocket to simulate the effect of the Sun's heat on the electronic systems. Four of these arrays were on the upper end of the Centaur, and two arrays were located near the RL-10 engines. Certain lamps could be turned on at different times to simulate the changing direction to the Sun.[458] (NASA C–1967–00180)

Image 215: The Vacuum Pump House contained the two Stokes piston pumps, which could remove 12.5 cubic feet of air per second during the roughing stage; a Roots rotary positive displacement pump, which could then remove 500 cubic feet per second, and ten 32-inch-diameter oil-diffusion pumps, which could remove 17,650 cubic feet of air per second.[359, 360] (NASA C–1962–60342)

Image 216: The SPC No. 1 control room was constructed underneath the former wind tunnel test section. It was built to replicate the Centaur controls at Cape Canaveral. This panel operated the tanking system for the Centaur 6A rocket in the vacuum chamber. There was extensive instrumentation for the tests, including 200 transducers, 18 landlines, and a television camera. (NASA C–1967–00193)

Electronics in the Space Environment

The effect of heat, both on the electronics and emanating from the electronics, was one of Lewis's primary concerns going into the Centaur tests. The liquid-hydrogen propellant had an extremely low boiling point, so any additional heat could cause problems. The electronics themselves could also heat up when exposed to the Sun and cause malfunctions. The simulated missions in Space Power Chamber (SPC) No. 1 were used to study these heat-transfer problems.

Since the launch time had not yet been determined, the Centaur systems were tested under a variety of temperature conditions ranging from 50 to 150 percent of what was actually expected on the mission. The researchers, who included Ralph Schmiedlin, Larry Ross, and Bob Turek, found no significant changes to the electronics at any of the levels, so the remaining tests were conducted at a constant radiation level.[461]

Image 217: NASA Lewis researcher John Povolny monitors telemetry equipment set up in the 8- by 6-Foot Supersonic Wind Tunnel Building for the Centaur environmental tests being run in SPC No. 1. This auxiliary area was used for inspection, calibration, and failure analysis of the guidance system and the inverter. Povolny would become the Chief of the Test Engineering Branch in the Centaur Program Office. (NASA C–1965–75077)

In the end the researchers determined that the electronics did not transfer heat to the spacecraft or propellant tanks. When not exposed to the Sun, the electronics packages cooled in the space environment. It remained uncertain, however, whether or not the solar radiation would cause problems with the electronic packages. Lewis researchers recommended that electrical systems for spacecraft be designed to run at the minimum necessary power level to avoid possible overheating. Solid-state electronics were chosen because they produced less heat.[462]

Of the many problems facing NASA Lewis researchers when they assumed control of the Centaur Program in 1962, the electrical inverter was one of the most daunting. The inverter produced power for Centaur's guidance and autopilot systems by converting direct current to alternating current. Although the device was vital to the Centaur's performance, little was known about its operation, use, or safety.

Image 218: Static inverter for Centaur guidance and autopilot systems. A more powerful off-the-shelf inverter was used after the first Centaur failure. After modifications at Lewis, the equipment was qualified in SPC No. 1. By 1965 it had successfully passed the design proof tests and was operating reliably. The inverter weighed 37 pounds and measured 10 by 16 inches. It was able to run for about 100 hours following a 12-hour prelaunch charge.[463] (NASA C–1964–70983)

After the first Centaur launch failed, a larger inverter was employed to better handle the power loads. A Lewis circuit specialist analyzed the inverter and modified it so that its components were within the mission's safety margins. Following successful operation during the SPC No. 1 vacuum tests, the updated inverter was removed from the Centaur critical list.[464]

Ensuring *Surveyor's* Survival

The 30 June 1964 launch of AC-3 was successful, but the rocket's guidance computer had been interrupted briefly because of the ejection of the nose fairing. Although this anomaly did not cause the overall Atlas/Centaur mission to fail, it was a concern for future launches. The nose fairing provided an aerodynamic shield for the payload, guidance system, and electronics package as the rocket traveled through the Earth's atmosphere. Upon entering space,

Image 219: On 5 August 1964, the first actual jettison test was run in SPC No. 1 at a pressure altitude of 70 miles. The deflector bulkheads below the payload were ripped from their bindings, and the tips of the shroud were broken off as the fairing halves slammed into the catcher pads.[465] (NASA C–1964–71125)

Image 220: Researchers in SPC No. 1 examine the nose cone for the Centaur/Surveyor spacecraft. (NASA C–1964–71093)

the thruster near the tip of the fairing forced the two pieces away from the space vehicle. It was essential that the *Surveyor's* mast and solar panels not be damaged during that brief jettison process.[466] The separation system had shown no difficulties during preflight tests in ambient temperatures and pressures. The Centaur team at Lewis felt that the conditions in space may have affected the system's two gaseous nitrogen bottles, which were used to activate the shroud's explosive bolts.[467]

Within a month of the incident, a Centaur fairing was obtained and installed in SPC No. 1—which was the only space tank in the country large enough to accommodate the hardware. The SPC tests, led by Jack Humphrey and Clarence Ross, sought to determine design faults in the AC-3 fairing and then flight-qualify the modified shroud for the upcoming AC-4 mission.

The two halves of the fiberglass fairing were mounted vertically to a platform at the opposite end of the chamber from the Centaur rocket used for the environmental tests. Aluminum pads were set up on either side to catch the fairing halves as they were jettisoned, and a myriad of high-speed cameras were installed to record the tests.

Image 221: Nitrogen bottle, thruster, and attachment plate in the tip of the Centaur/Surveyor nose cone. This thruster fired as the explosive bolts securing the fairing were released, thus pushing the 1,000-pound fairing halves away from one another and the Centaur. (NASA C–1965–00604)

Image 222: Jack Humphrey, project manager for the Surveyor nose-cone tests, inspects half of the jettisoned shroud inside SPC No. 1 in 1964. Humphrey had been at the center since 1943. He and his team had determined that the new design had enough clearance between the deflector bulkhead and the packages. (NASA C–1964–73201)

In late July the shroud was heavily damaged during the very first run conducted in altitude conditions. The force of the thruster caused the tips of the fairing to break as they hit the catcher pads. The excess force would likely damage the *Surveyor* on an actual launch. The team regrouped and obtained a new shroud. The internal bulkhead was redesigned, and a new attachment fixture was installed for the thruster. The test setup was modified as well. The platform was moved further from the Centaur at the other end of the chamber, and one of the catcher pads was replaced by a large net. This allowed the researchers to study the entire path of one of the jettisoned halves.[468]

Over the course of 11 ensuing runs, the redesigned bulkhead was tweaked and retested. Though there was slight damage at times, the flight-worthiness of the new fairing was validated by the final jettison on 24 November 1964. The Centaur managers were confident that the thruster devices would jettison the fairing without damaging the payload. Lewis researchers recommended that these separation tests must be conducted in a vacuum environment.[469]

Just over two weeks later, AC-4 successfully launched a mock-up *Surveyor* spacecraft into orbit. It was the first Centaur mission to have an error-free shroud jettison.[470] During the coast phase of the flight, however, the Centaur spun out of control. This would lead to a whole new vein of propellant management studies at NASA Lewis. The 3 March 1965 AC-5 was an even more spectacular failure. The Atlas booster's engines failed seconds into the launch, and the rocket exploded on the launch pad.[471]

Centaur's Long Days

NASA Lewis Director Abe Silverstein personally oversaw the program during its initial years and hand-picked some of the center's brightest people to run it. The staff was a combination of old NACA veterans and young engineers right out of university. Larry Ross was one of the latter. He joined the program in June 1963 after completing his undergraduate degree at Manhattan College. Ross accepted the position primarily because Lewis offered assistance with graduate degrees, but he found his involvement with the Centaur Program to be challenging and rewarding. The following year he met his future wife while working in the Blockhouse. Ross spent the next 30 years at Lewis and continues to use his Centaur experience to advise the current Ares engineers.[472]

Dave Gabriel was Centaur Project Manager during the initial years, with Cary Nettles, Russ Dunbar, and Ed Jonash as his assistants. Former Centaur lawyer, Harlan Simon referred to these men as "absolute giants."

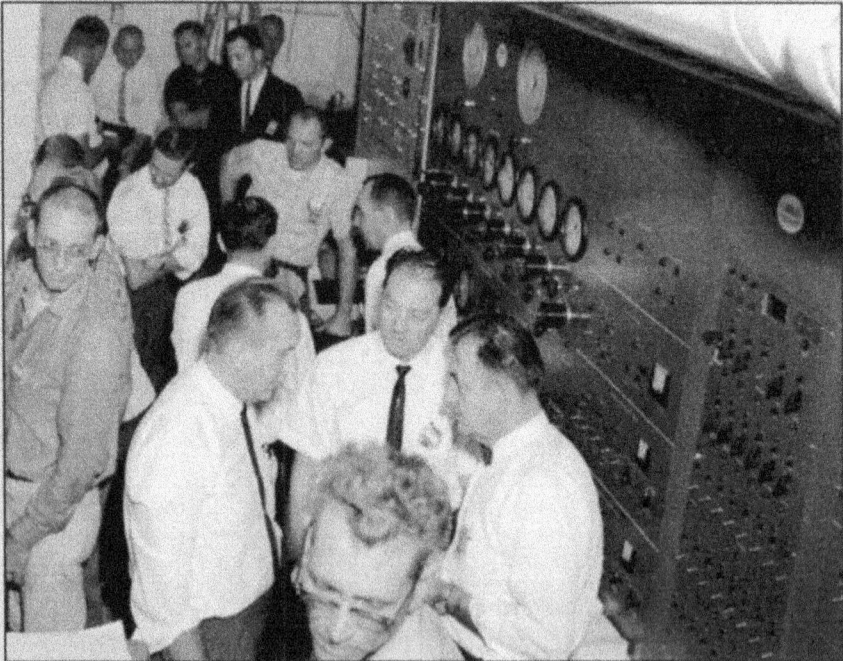

Image 223: Lewis researchers and General Dynamics technicians crowd into the SPC No. 2 control room to prepare for Atlas/Centaur retrorocket tests. The cramped control room was in the former Altitude Wind Tunnel control room built in 1943. (NASA C–1963–66372)

During a 1985 interview with Virginia Dawson, Simon emphasized that, besides being technically savvy, they were also excellent managers. "Whenever somebody came in with a statement or conclusion which in their minds was obviously wrong, they would approach it from an educational standpoint. They would ask questions and finally they would bring the presenter around to the point where he saw that his recommendation or presentation was not feasible, was not correct."[473]

There was tremendous pressure on the Lewis staff to make the program succeed. The team, constantly driven by upcoming launch dates, invested an extraordinary amount of time in the program. Overnight trips between work days to California or Florida were routine for many. Simon claimed, "They worked seven days a week in five days. Nine days in a week sometimes. They had boundless energy.... These people gave their lives essentially."[474]

Image 224: A meeting of Centaur Program managers. Standing, Cary Nettles; in bowtie, Al LeRoy; to LeRoy's left, Jeff Essary. During the program's initial years, the group doggedly worked through problems. It was not uncommon for meetings to last 6 to 8 hours. (NASA MPD–538)

The success was not without its casualties, however. In 1965 Gabriel took a two-year leave of NASA. The stress of the program, along with the heavy involvement of Silverstein, had taken its toll. Former mission analyst, Joe Nieberding, explained the sacrifices made by the Centaur staff. "It's tough to balance the family and the personal situations against a tremendous program like this. When people are this dedicated, the family sometimes loses."[475] Divorces were the most common result, but there were also break-downs and shortened lives.[476]

The experience and high visibility of the Centaur Program, however, benefited the careers of many young engineers. Former NASA Chief Historian Sylvia Fries noted that 90 percent of the Apollo-era engineers had entered the management ranks by the end of the 1980s. Salaries plateaued for those engineers who resisted the transition.[477] Some of the former Centaur engineers would move into the upper echelon of the Lewis organization. This is exemplified by the progression of Andy Stofan and Larry Ross from inexperienced engineers right out of university in the early 1960s to NASA center directors in the 1980s. Ross recalled in 2007 that early in his career, SPC Manager Ralph Schmiedlin had correctly predicted that Ross would one day lead the center.[478]

Centaur Turns the Corner

The fiery launch pad failure of AC-5 was caused by the booster engines. In response, modifications were made to Centaur that required the requalification of its nose cone prior to the AC-6 direct-ascent launch scheduled for August 1965. NASA was becoming anxious. AC-5 was supposed to place the first *Surveyor* model into orbit. This would now be the responsibility of the sixth of eight development launches.[479]

Potentially dangerous shrapnel produced during the AC-4 nose-cone tests in the SPC had led to the redesign of the fairing. The new shroud design, which incorporated approximately 25 new components, required the requalification of the original payload clearance envelope. The Centaur's forward bulkhead would also have to be reassessed to ensure that the new tanks would hold up during the shroud jettison.[480]

Image 225: Nose-cone separation test in SPC No. 1 for the upcoming AC-6 mission, which would place the first Surveyor model into orbit. The 2,000-pound model was used as an instrumented dynamic payload simulation. (NASA C–1965–00725)

The SPC tests in the early summer were once again conducted by Jack Humphrey along with Charles Eastwood. The SPC No. 1 setup and test conditions were almost identical to the AC-4 shroud tests the previous fall. The envelope between the thermal bulkhead struts and the *Surveyor* had to be altered because of interference during the separation process. The redesigned shroud was approved in July after a series of runs in the SPC.[481] AC-6, launched from Cape Canaveral on 11 August 1965, successfully placed the *Surveyor* model into an elliptical Earth orbit. The mission was a pivotal moment in the program's history. It not only restored the nation's confidence in the Centaur's capabilities but showed that failure could be turned into success.

Image 226: Cary Nettles in the SPC shop at the foot of the Centaur 6A rocket. (1963) (NASA MPD–538)

Nettles told the Centaur team in 1966, "I believe a turning point in the project also came at our greatest failure with the flight of AC-5, which has been rather generously described in the NASA Project Summary as a suborbital flight. Looking back on this now, from kicking your asses around at [launch pad] 36A in March, to the successful launch in AC-6 just five months later from a new complex, was a fine accomplishment that really made professionals out of all of you. This really turned our greatest failure to the source, I think, of our greatest strength."[482]

Another Chance at Two-Burns

The Centaur team did not have time to enjoy the success of AC-6, though. They had already been looking ahead to the crucial AC-8 mission scheduled for April 1966.[§] AC-8 would be the second attempt to restart the RL-10 engines in space. The previous attempt on AC-4 had failed when the hydrogen

---

[§]The Centaur missions were assigned numbers as the missions were conceived. Between mission conception and the actual launch, priorities sometime shifted, so the launch numbers were not always in order. AC-7 would be launched on 20 Sept. 1966 after both AC-8 and AC-10.

Image 227: Left to right: Ben Dastoli and Larry Ross prepare for a test in the SPC No. 1 control room. (1964) (NASA C–2008–01424)

propellant sloshed forward and out the tank's vents after the first engine firing. It was up to the NASA Lewis engineers to come up with a solution to the sloshing problem.

AC-8 would also use an updated electronics package that had to be verified in the SPC No. 1 space tank. The electrical reharnessing of the Centaur 6A already in the chamber took several months. The intensive procedure was slowed down by delays in receiving the upgraded hardware from General Dynamics. The company was simultaneously trying to supply the launch teams at Cape Canaveral, their own test engineers, and Lewis. Delays in the production of hardware were common.[483]

After the AC-4 mission, a new C-band transponder, or beacon, for the Centaur's tracking system was incorporated in the design. The new transponder was tested in SPC No. 1 approximately 30 times over the next year without problem. A faulty seal, however, caused the transponder to fail on AC-6, which crippled the entire tracking system. During the lead-up to AC-8, Lewis engineers were able to recreate the failure in the space chamber. The transponder that was to be used for the test was faulty, so Larry Ross personally returned the device to the manufacturer in Phoenix. It was repaired overnight, flown back to Lewis, and installed on the Centaur the next day.[484]

Although a seal caused the failure on AC-6, Lewis engineers found that the larger problem was the pressurization of electronics boxes in space. Many of the boxes were designed to operate at specific pressures. The engineers connected the boxes to a purge system so that they could analyze the effect of pressure loss on the performance of the electronics. They found that the pressurization of the electronics was not needed and could actually cause failures. This resulted in an industrywide redesign of electronics packages.[485] The team recommended that the electronic systems be designed to act independently and that a verification of the transponder seal be added to the standard series of preflight readiness checks.[486]

Image 228: The Centaur 6A was initially harnessed with electronics and other equipment designed for the AC-4 mission. Afterward, the rocket was reharnessed with a new configuration for the AC-8 flight. This photograph shows the guidance packages that were mounted near the forward end of the rocket. Centaur's inertial guidance system consisted of five components: an inertial platform, platform electronics, a pulse rebalance with power supply, a navigational computer, and a signal conditioner. (NASA C–1969–02623)

During the Centaur's SPC system tests conducted for AC-4, a metal shard pierced the insulation between the heat sink and the overload-sensing shunt. This caused the static inverter to fail during the test. The inverter created power for the Centaur's guidance and autopilot systems. Lewis researchers felt that this warranted a detailed investigation, since a similar event would cause an actual mission to fail.[487]

The researchers sought to replicate the failure from the original test in SPC No. 1 at a pressure altitude of 390,000 feet. Despite a couple of early interruptions, the inverter had operated successfully in the chamber for a total of 79 minutes. It was found that during the repeated temperature fluctuations of the original tests, the shard forced open the shunt, which caused the inverter to fail. The researchers felt that this was just one instance of a poorly manufactured component that should have been analyzed before use. They recommended that Centaur systems should go under repeated environmental tests individually before the entire vehicle was tested.[488]

The SPC No. 1 environmental tests played a key role in the U.S. lunar program. This series of studies in the SPC proved that the Centaur's electronics systems could perform during a two-burn flight in a space environment. General Dynamics's original design was sound, and the few problems that were found were rectified before the *Surveyor* flights.[489]

Managing Liquid Hydrogen

Centaur's use of liquid hydrogen as a propellant posed a number of technical difficulties. The cryogenic liquid vaporized at a very low temperature, so the propellant tanks had to be designed with vents to prevent an explosion. Solar energy coming directly from the Sun and reflected off the Earth's atmosphere caused some of the liquid hydrogen to boil off. In addition, because the Centaur's engines had to be restarted in space, the liquid hydrogen and its liquid-oxygen oxidizer had to be stabilized within their tanks during the coast period between engine firings. Otherwise the propellant might not be in the correct position when it came time for the engines to restart.[490]

The AC-4 and AC-8 missions in December 1964 and April 1966 were designed to be two-burn missions. The focus would be on managing the behavior of the propellant within the tank so that the engines could be restarted.[491] When the first engine burn ended on the AC-4 flight, the liquid hydrogen sloshed forward, resulting in the venting of some of the hydrogen in liquid form rather than gas. Normally the venting was even and nonpropulsive, but the forces generated when liquid was vented were more than the guidance system could overcome.

Image 229: Centaur's cryogenic liquid-hydrogen system is tested in SPC No. 2 prior to the AC-4 mission. The cryogenic liquid hydrogen used by Centaur had a boiling point of −423°F. During the Centaur's coast period, some of the fuel would vaporize and would have to be expelled via vents. The AC-4 system contained a venturi pump that measured the gaseous hydrogen flow from the tank through the standpipe. The pressure level was controlled by a vent valve. After the nose cone was ejected, the gas could be vented through pressure-filled exits. (NASA C–1964–72930)

Image 230: Setup for qualification tests of the new hydrogen venting system in SPC No. 2 prior to the AC-8 mission. The hydrogen gases were drawn up into a small chamber, which then discharged through symmetrical vents. The nozzles were installed away from the spacecraft to dampen their flow effects. Shields were placed below the inlet to keep the vent lines free of liquid hydrogen.[492] (NASA C–1965–03932)

The motion of the liquid hydrogen prevented the vehicle from maintaining its balance, and the uneven venting skewed the Centaur's trajectory, which caused the loss of even more liquid hydrogen. Approximately 90 percent of the liquid hydrogen was lost during the coast phase, and the engines could not be restarted. The tumbling spacecraft fell back into the Earth's atmosphere.[493]

Although NASA Lewis's Weightlessness Analysis Sounding Probe and Aerobee launches had previously studied the behavior of liquid hydrogen on scaled models, the failed AC-4 flight revealed the unique problems created by the forces associated with full-size propellant systems. Lewis researchers undertook a series of propellant management studies that resulted in several modifications for the AC-8 flight. The vent system was completely redesigned, energy dissipaters were added, and a baffle was inserted in the hydrogen tank to prevent sloshing.

The new AC-8 vent system drew the hydrogen gases up into a small chamber, from which they were discharged through symmetrical vents. The even distribution of this thrust was perfected with numerous ground tests. In December 1965 the system was qualified during an extensive series of runs in SPC No. 2.[494]

On 8 April 1966, the AC-8 became the first Centaur to restart its engines in space. The propellant was successfully managed, and the off-gasses were expelled without altering the rocket's trajectory.[495] The mission was even a larger achievement than the AC-6 success. The Centaur's complex two-burn capacity had finally proven itself. For the first time, NASA and Congress felt confident in the Centaur's abilities.

Moonshot

AC-8 was the Lewis team's biggest success to date, but as usual, there was little time to celebrate. The next launch, less than two months later, would be its first attempt with a genuine payload. On 30 May 1966, the AC-10 vehicle lifted off from Cape Canaveral carrying the *Surveyor 1* spacecraft. The Centaur stage performed perfectly, and *Surveyor* was on its way. Three days later, as the NASA Lewis Centaur team watched on television at Lewis's Guerin House, the *Surveyor* began sending back images from the surface of the Moon. Cary Nettles recalled later, "[AC-10] was the ultimate climax of over four years of intensive work. I cannot really explain the elation that I felt in this personal triumph, that everything in the flight was perfect.[496]

Image 231: The AC-10, carrying the Surveyor 1 spacecraft, lifts off from Pad 36A on 30 May 1966. On 2 June, the Surveyor 1 became the first spacecraft to land on the Moon. It was Centaur's second successful two-burn mission and one of NASA Lewis's greatest accomplishments. (NASA GPN–2000–000617)

NASA Deputy Administrator Robert Seamans immediately wrote Abe Silverstein, "This total performance justifies the faith which was registered approximately three and one-half years ago in both the Centaur concept and the competency of the Lewis Research Center." He added, "The achievement of the Centaur organization, both government and contractor, is particularly striking when this technical difficulty and complexity of the Centaur development must have appeared to be nearly insurmountable. However, the outstanding technical confidence of the people and the dedication which they have given to the Centaur Program seems to have reduced all of these difficulties to manageable size. These people have shown they are truly professional engineers and managers."[497]

Ross later pointed out, "We had a remarkably competent team of mechanics and technicians. They were just the best…. I call them 'the can-do people.' Nothing would have happened by way of understanding the systems contributing to the reliability [of Centaur] without Silverstein's vision and the can-do team. The engineers were important, but the mechanics and technicians really made it happen."[498]

Centaur was Lewis's first attempt at a large developmental program and its first hands-on experience with spaceflight systems. It was a triumph. The *Surveyors* would prove the ability to soft land on the Moon, explore landing sites for the Apollo missions, and perform geological studies. On 7 October, a NASA Group Achievement Award was awarded to 123 members of the Lewis Centaur Program staff.[499] Four years of intense work by the NASA Lewis Center Director, management, engineers, technicians, and mechanics had paid off.

Yet at the ceremony, Nettles turned to the future. He said, "In the broadest sense, we're not standing here today at the completion of Centaur's development. It's not the end or even the beginning of the end. This is rather, I think, the end of the beginning…a chance to transform Centaur from a successful launch vehicle for *Surveyor* into an effective general-purpose upper stage for a variety of high-energy missions."[500]

Endnotes for Chapter 10

446. Kim McQuaid, "The Space Age at the Grass Roots: NASA in Cleveland, 1958–1990, 2007," NASA Glenn History Collection, Cleveland, OH, pp. 12–14.
447. McQuaid, "The Space Age at the Grass Roots."
448. Bruce Byers. *Destination Moon: A History of the Lunar Program* (Washington, DC: NASA TM X–3487, 1977), chap. 2.
449. "Surveyor 1: NSSDC ID:1966–045A," NASA's National Space Science Data Center (7 September 2006), *http://nssdc.gsfc.nasa.gov/database/MasterCatalog?sc=1966-045A* (accessed 8 October 2009).
450. "Surveyor 1: NSSDC ID:1966–045A."
451. Interview with Art Zimmerman, conducted by Virginia Dawson and Mark Bowles, 5 August 1999, NASA Glenn History Collection, Cleveland, OH.
452. Ralph Schmiedlin, et al., *Flight Simulation Tests of a Centaur Vehicle in a Space Chamber* (Washington, DC: NASA TM–X–1929, 1970), p. 4.
453. "Surveyor 1: NSSDC ID:1966–045A."
454. Cary Nettles, "Talk at NASA Group Achievement Award Ceremony at NASA Lewis Research Center," 1 December 1966, Cleveland, OH.
455. Schmiedlin et al., *Flight Simulation Tests,* p. 1.
456. Schmiedlin et al., *Flight Simulation Tests.*
457. Schmiedlin et al., *Flight Simulation Tests,* p. 14.
458. Schmiedlin et al., *Flight Simulation Tests,* pp. 12 and 35.
459. Schmiedlin et al., *Flight Simulation Tests,* pp. 35 and 39.
460. Jack Humphreys, *Centaur AC–4 Nose Jettison Tests* (Washington, DC: NASA TM–X–52154, 1966), p. 3.
461. Schmiedlin et al., *Flight Simulation Tests,* pp. 1 and 16–17 and 22.
462. Schmiedlin et al., *Flight Simulation Tests,* pp. 1 and 16–17 and 22.
463. Baddour, Maurice, *Centaur Electrical System Problems, Related Work, and Solutions* (Washington, DC: NASA TM–X–52105, 1965), p. 5.
464. Baddour, *Centaur Electrical System,* p. 2.
465. Humphreys, *Centaur AC–4 Nose Jettison Tests,* p. 6.
466. R.L. Radcliffe, "Surveyor Nose Fairing Mode Survey," 28 April 1964, Evaluation Test Report No. 55A3291, NASA Glenn History Collection, Cleveland, OH, p. 4.
467. Humphreys, *Centaur AC–4 Nose Jettison Tests,* p. 1.
468. Humphreys, *Centaur AC–4 Nose Jettison Tests,* pp. 4–6.
469. Humphreys, *Centaur AC–4 Nose Jettison Tests,* p. 7.
470. John Povolny, *Space Simulation and Full-Scale Testing in a Converted Facility* (Washington, DC: NASA N66 35913, 1970), p. 53.
471. Virginia Dawson and Mark Bowles, *Taming Liquid Hydrogen: The Centaur Upper Stage Rocket, 1958–2002* (Washington, DC: NASA SP–4230, 2004), p. 84.
472. Interview with Larry Ross, conducted by Bob Arrighi, 1 March 2007, NASA Glenn History Collection, Cleveland, OH.
473. Interview with Harlon Simon, conducted by Virginia Dawson, 20 March 1985, NASA Glenn History Collection, Cleveland, OH.
474. Simon interview, conducted by Dawson, 20 March 1985.

475. Interview with Joe Nieberding and Robert Gray, conducted by Virginia Dawson, 9 November 1999, NASA Glenn History Collection, Cleveland, OH.
476. Simon interview, conducted by Dawson, 20 March 1985.
477. Sylvia Doughty Fries, *NASA Engineers and the Age of Apollo* (Washington, DC: NASA SP–4104, 1992), p. 137.
478. Larry Ross to Bob Arrighi, "AWT Chapters 10–11," 16 June 2008, NASA Glenn History Collection, Cleveland, OH.
479. "Success Scored by Atlas-Centaur 6 Flight," *Lewis News* (20 August 1965), p. 1.
480. Jack Humphreys and Charles Eastwood, *Centaur AC–6 Nose Fairing Separation Tests* (Washington, DC: NASA TM–X–52290, 1965), p. 1.
481. Humphreys and Eastwood, *Centaur AC–6 Nose Fairing.*
482. Nettles, "Talk at NASA Group Achievement Award Ceremony."
483. Interview with Larry Ross, conducted by Bob Arrighi, 1 March 2007, NASA Glenn History Collection, Cleveland, OH.
484. Ross interview, conducted by Arrighi, 1 March 2007.
485. Ross interview, conducted by Arrighi, 1 March 2007.
486. Schmiedlin et al., *Flight Simulation Tests,* pp. 15 and 21.
487. Schmiedlin, Ralph and B.J. Dastoli, *Centaur Solid State Inverter Failure During Simulated Flight in an Environmental Space Chamber* (Washington DC: NASA TM–X–52336, 1967), p. 1.
488. Schmiedlin and Dastoli, *Centaur Solid State Inverter,* pp. 1–2.
489. Schmiedlin et al., *Flight Simulation Tests,* p. 1.
490. Lacovic, Raymond, et al., *Management of Cryogenic Propellants in a Full-Scale Orbiting Space Vehicle* (Cleveland, OH: NASA TN D–4571, 1968), p. 14.
491. Groesbeck, William, *"Design of Coast-Phase Propellant Management System for Two-Burn Atlas-Centaur Flight AC–8* (Washington DC: NASA TM–X–1318, 1966), p. 2.
492. Lacovic, et al., *Management of Cryogenic Propellants,* p. 15.
493. Lacovic, et al., *Management of Cryogenic Propellants,* pp. 1 and 15.
494. Groesbeck, *Design of Coast-Phase Propellant Management System,* pp. 2 and 15.
495. Lacovic, et al., *Management of Cryogenic Propellants,* p. 1.
496. Cary Nettles to Virginia Dawson, "Centaur Notes," 8 May 2002, NASA Glenn History Collection, Cleveland, OH.
497. Seamons to Silverstein, "Atlas Centaur Success."
498. Ross interview, conducted by Arrighi, 1 March 2007.
499. "Achievement Awards Go To 146 Centaur Project Workers," *Lewis News* (9 December 1966), p. 1.
500. Nettles, "Talk at NASA Group Achievement Award Ceremony."

Image 232: Lewis engineers stand below the net used to catch the OAO-1 fairing in SPC No. 2. Following the Surveyor launches, Centaur was used to carry increasingly large payloads, such as the OAO satellites. The Space Power Chambers was used to test the shroud jettison system for many of these new missions. (NASA C–1965–01628)

Chapter 11

Alter Destiny | Centaur's Big Payloads (1965–1975)

"They amazed me. I didn't think Lewis, being a propulsion center, would have the smarts to do some of that stuff," recalled former NASA Lewis Research Center engineer and Deputy Associate Administrator for Space Science and Applications, Ed Cortright.[501] Lewis's rocket programs had matured quickly in the early 1960s. The old engine lab quickly became a leader in launch vehicles. Cary Nettles recalled, "The whole aerospace community suddenly took notice that Lewis was a first-line contender in the space business."[502] The staff had physically moved from the temporary cinderblock offices in the hangar to the brand new Development Engineering Building with a control room linked directly to Cape Kennedy. From there the staff could monitor and back up the Lewis launch team at the Cape. The staff also moved professionally from inexperience to expertise in the fields of spaceflight systems, payload integration, and launching.

The missions expanded, too. After Centaur's achievements with *Surveyor,* its future was promising. The vehicle underwent numerous technical adjustments and was ready for larger payloads and more complex missions. Not only was it ready, but it was needed. The Agena rocket was the only other comparable vehicle in the NASA stable, and it had a limited lift capability. Centaur was the Agency's only option for launching heavy payloads.[503] The larger payloads required larger fairings and special modifications to integrate the satellites into the launch vehicle. Over the next 10 years, the Space Power Chambers (SPC) would be critical for testing the new fairing configurations.

Lewis's Other Rocket—Agena

Lewis also was responsible for the Agena second-stage rocket program. Like the Centaur, it was an upper-stage space tug with a two-burn capability. Its single Bell 8096 engine produced considerably less thrust than Centaur's two liquid-hydrogen RL-10s, however. Agena performed numerous important missions that included the first closeup photographs of the Moon, Mars, and Venus.

Between its debut in 1959 and its transfer to Lewis in January 1963, the Agena rocket had put over 80 launches under its belt. Since Agena was an established technology, Lewis engineers did not have to overhaul it as they did Centaur. The first Lewis launch was a Thor/Agena carrying the Echo 2

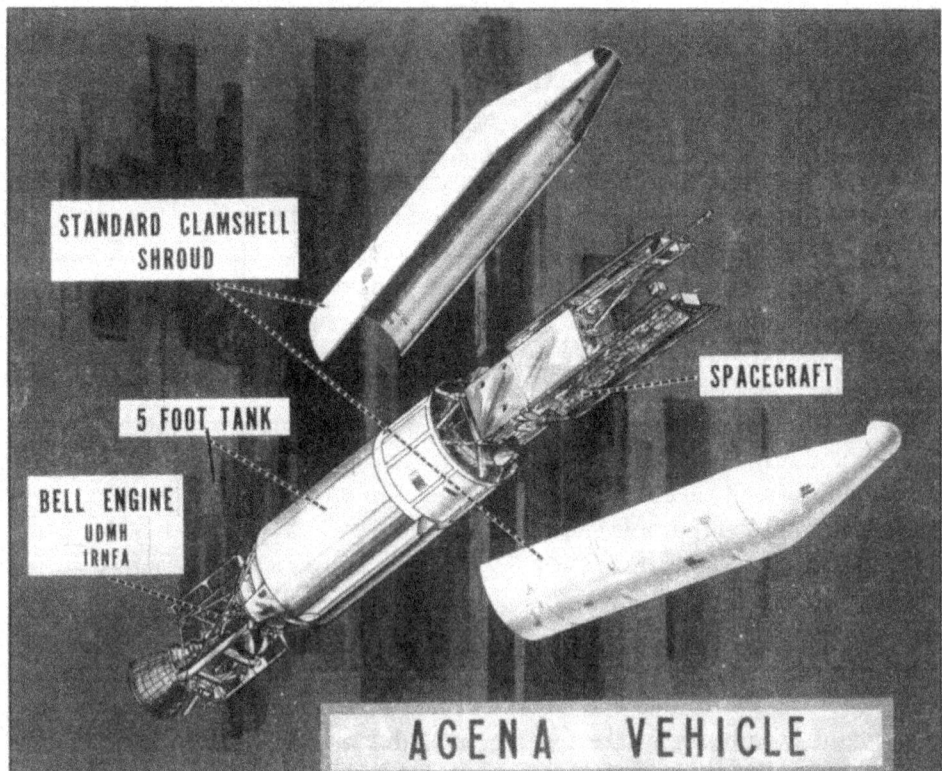

Image 233: The versatile upper-stage Agena rocket employed a single 16,000-pound-thrust engine that could be shut down and restarted in flight. Agena and its payload reached space on Atlas, Thor, and Titan boosters. Following staging, the first burn maneuvered the Agena into an Earth-oriented parking orbit. The second burn was geared to the particular mission— either an elliptical Earth orbit or trajectories toward the Moon or planets. (NASA C–1968– 01439)

Image 234: Lewis Agena team whiteboards for the 21 March 1965 Ranger C launch. The boards were used to track flight and countdown events. For each Agena launch, Lewis defined the launch vehicle requirements; acquired, tested, and integrated the vehicle; prepared for the launch; and oversaw the launch until the stage was at its correct flight trajectory.[504] (NASA C–1965–74460)

satellite balloon. The *Ranger 6* launch followed five days later.[505] During the seven years that Lewis managed the Agena Program, 27 of 30 launches were successful. Agena's heaviest payload attempt would be the 3,900-pound Orbiting Astronomical Observatory 1 (OAO-1) satellite in April 1966.[506]

Observatories in Space

The OAO Program, begun at the NASA Goddard Space Flight Center in 1960, consisted of a series of four increasingly heavy and complex space observation satellites, direct predecessors of the Hubble Telescope. The satellites were equipped with powerful telescopes to study and retrieve ultraviolet data on specific stars and galaxies. In-depth observations were not possible from Earth-bound telescopes because of the filtering and distortion of the atmosphere. NASA hoped that the ultraviolet data would help researchers determine the age of certain stars. The telescopes required a large stable platform so that they could focus on dim and distant stars for long periods of time.[507]

The process of placing the OAO-1 in its 500-mile-altitude Earth orbit was expensive, so the satellite was designed for a lengthy life.[508] Years of preparation were involved, and a great deal was riding on a successful launch. Goddard selected the Atlas/Agena D for the task despite the fact that the rocket had not previously borne that much weight. Lewis would be responsible for integrating the payload with the Agena, testing the setup, and launching the vehicle.[509]

The OAO-1 satellite was wider in diameter than the Agena stage, so a new three-section clamshell shroud was created to enclose both the satellite and the Agena. This new shroud would be qualified in SPC No. 2 under the supervision of William Prati and OAO Project Engineer Richard Geye. In June 1965, Lewis technicians began setting up the large fairing inside the chamber over models of the OAO-1 and Agena.

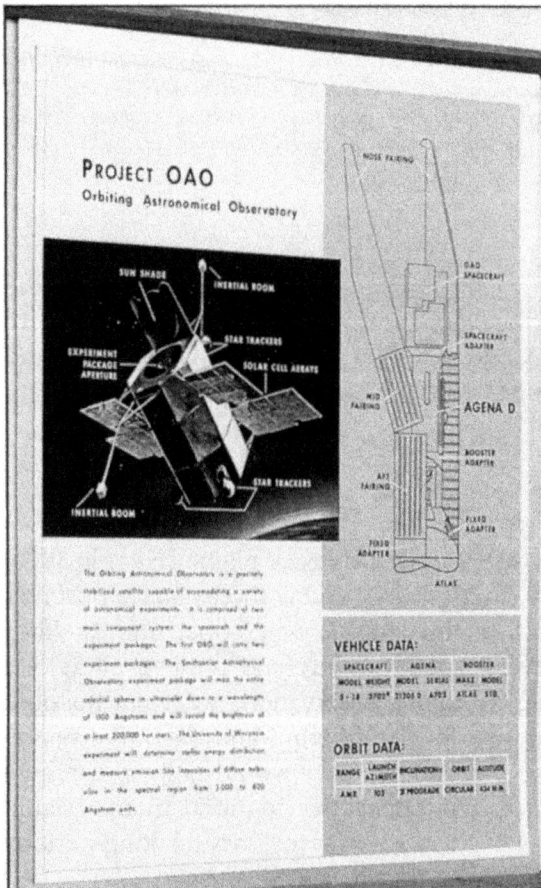

Image 235: The text on this 1965 poster states: "The OAO is a precisely stabilized satellite capable of accommodating a variety of astronomical experiments. It is comprised of two main component systems: the spacecraft and the experiment packages. The first OAO will carry two experiment packages. The Smithsonian Astrophysical Observatory experiment package will map the entire celestial sphere in ultraviolet down to a wavelength of 1100 Angstroms and will record the brightness of at least 200,000 hot stars. The University of Wisconsin experiment will determine stellar energy distribution and measure emission line intensities of diffuse nebulae in the spectral region from 3000 to 800 Angstrom units." (NASA C–1965–74461)

Image 236: Lewis engineers work on the OAO-1 test setup in SPC No. 2. One-half of the large clamshell nose fairing is removed in this photograph, revealing the 10-foot-long, 4-foot-diameter OAO-1 payload. The solar panels are folded flat against the satellite as they would be during a launch. A platform elevator with an 11-foot inside diameter was built around the setup to allow access to all areas of the shroud. (NASA C–1965–01461)

Three jettison tests were run in July and the first week of August at a simulated altitude of 20 miles. For these studies, only one-half of the fairing was ejected. A large net was stretched horizontally over the chamber floor to catch the jettisoned shroud. Accelerometers on the model and shroud provided William Prati and Richard Geye with data they could use during the actual launch to verify a successful separation.[510]

The launch from Cape Canaveral in April 1966 was trying for the Lewis team, however. Geye recalled, "On the first attempt we scrubbed the mission because of instrumentation problems. Then it was hot fired. The third time a tornado in the area blew out our power supply. Another hot firing followed. We finally made it on the fifth try."[511] The launch and separation during the fifth try on 8 April 1966 went smoothly. The OAO-1 satellite itself, however, failed after only 90 minutes when overarcing in the star trackers caused a battery failure.[512]

Photo Essay 7:
Space Power Chambers' Busy Schedule

Image 237: The setup for a shroud separation test prior to Centaur's Orbiting Astronomical Observatory (OAO-1) launch. The clamshell shroud consisted of three sections that enclosed both the Agena and OAO—a fiberglass nose fairing and aluminum mid and aft fairings. The upper two fairings separated when the Atlas engines stopped, and the aft fairing fell away with the Atlas upon separation from the upper stages.[513] Grumman Aerospace Corporation built the OAO-1, Lockheed the Agena, and General Dynamics/Convair the shroud and separation system. (NASA C–1965–02079)

Image 238: A mechanic installs the new platform elevator in SPC No. 2 for a series of Centaur/OAO shroud tests. (NASA C–1965–00138)

Image 239: The viewing platform alongside the former tunnel test section was converted into a shop area. The high bay is to the left of the platform, and the test section is to the right. (NASA C–1965–00378)

Image 240: The Weightlessness Analysis Sounding Probe is prepared for testing in SPC No. 2 in August 1964. The probe was a two-stage sounding rocket designed by Harold Gold and NASA Lewis's Spacecraft Technology Division to examine ways to control liquid hydrogen during the periods between rocket firing. The rocket would carry a transparent scale model of the S-II fuel tank and television cameras to film the behavior of the propellant during flight.[514] *(NASA C–1964–71575)*

Image 241. The Weightlessness Analysis Sounding Probe sounding rocket is tested in SPC No. 1 in April 1966. During the 7 June 1966 sounding rocket launch off of Wallops Island, the propellant was purposely sloshed by a thruster on the side of the tank. The tank contained a baffle ring that quickly settled the sloshing fuel.[515] The 862-pound rocket reached an altitude of 250,000 feet before freefalling back to Earth. The almost 7 minutes of microgravity during freefall provided researchers with enough data to launch the first orbital Saturn IB one month later.[516] (NASA C–1966–01838)

Image 242: One-half of an Atlas Launcher set up inside the high bay of the SPC facility. On the launch stand, the Atlas was tethered by two pins that were pulled out by the booster engine's thrust. Friction caused the pins to pull at different speeds, resulting in asymmetrical drag on the Atlas during its launch. The high bay was used for this test because of its thick concrete base. A series of tests conducted with the launcher's engines and hydraulics activated resulted in a resolution of the problem. (NASA G–1966–01269)

Image 243: In late 1965, tests of deceleration pellets for the new Zero Gravity Facility, which was about to come online, were conducted in SPC No. 2. A 5-foot-diameter, 20-plus-foot-tall deceleration stand was mounted vertically in the chamber floor and filled with polystyrene pellets. Objects were dropped through a penetration in the roof into the decelerant to determine which types of pellets worked best. (NASA C–1966–01954)

Centaur Moves Forward

Cary Nettles recalled making a presentation in the mid-1960s to a group from the new telecommunications group COMSAT about Centaur's payload capabilities. He told Virginia Dawson in 2002 that this meeting led directly to many of Centaur's communications satellite missions in the 1970s.[517] Centaur was already selected for the 1969 *Mariner* flights to Mars when, in January 1967, it assumed responsibility for three missions previously slated for the Agena—the next two Advanced Technology Satellites and OAO-2.[518]

The 4,436-pound OAO-2, the largest payload ever attempted on an Atlas/Centaur, was scheduled to be launched in late 1968 on Atlas/Centaur 16 (AC-16). The payload weight was redistributed after the failure of the first OAO satellite. The unprecedented size of OAO-2 forced NASA Lewis engineers to use a longer Agena shroud and a transition adapter on the Centaur vehicle. The basic Agena/OAO-1 separation system and fairing remained, but the fairing included a cylindrical section of the Centaur/*Surveyor* fairing and a new adapter to fix the fairing to the booster.[519] The modified shroud for the OAO-2 mission was 18 feet longer than the *Surveyor* nose fairing and was jettisoned by a mechanical spring rather than by a gas thruster system.[520, 521]

Image 244: A model of the OAO-2 observation satellite is transported through the former AWT test section and into SPC No. 2. OAO-2 contained two new experiments installed at opposite ends of the satellite. The Smithsonian Astrophysical Laboratory's experiment consisted of four telescopes capable of examining over 700 stars every day. The University of Wisconsin's study employed seven telescopes to study single stars for longer durations to determine the chemical composition, temperature, and pressure.[522] (NASA C–1968–01709)

Lewis's Launch Vehicle group was responsible for the compatibility of the launch vehicle and satellite. They determined if adapters or extensions were needed to ensure proper shroud clearance. The satellite manufacturer developed a "control envelope" to identify an area outside of the spacecraft's path. The Lewis team studied this envelope to ensure that neither spacecraft motion nor the shroud jettison interfered. After any conflicts were resolved and the payload was encapsulated, a final x ray of the spacecraft was taken to make sure that the payload was properly centered within the shroud.[523]

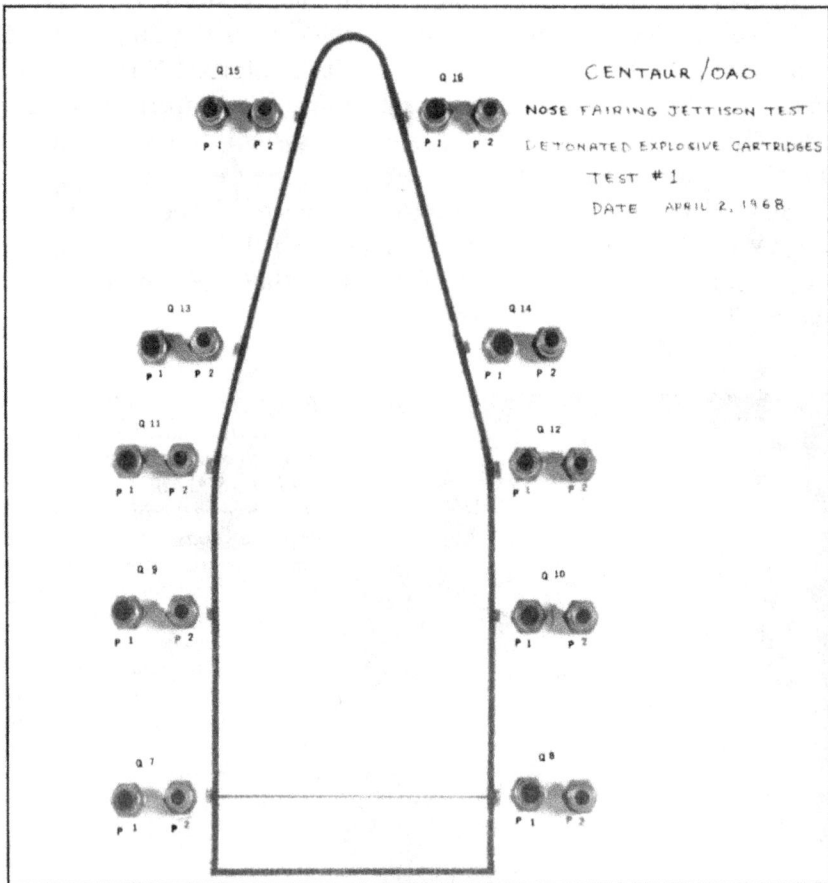

Image 245: Panel showing a diagram of the OAO-2 fairing and its latches. There were 10 latches along the vertical splitlines and 6 around the aft circumferential joint. Each latch was held by a single explosive nut, and each nut had two redundant explosive cartridges. The cartridges were activated by a mission programmer as the Centaur engines were fired. The tip of the cone was joined by the new mechanical spring thruster.[524] (NASA C–1968–01621)

As part of this process, Lewis researchers sought to qualify the new fairing with a series of jettison tests in SPC No. 2. Three tests were successfully run in April 1968 at a simulated altitude of 90,000 feet. The shroud's structural integrity held up, the fairing halves performed well, and the entire separation system worked in a simulated space environment.[525, 526]

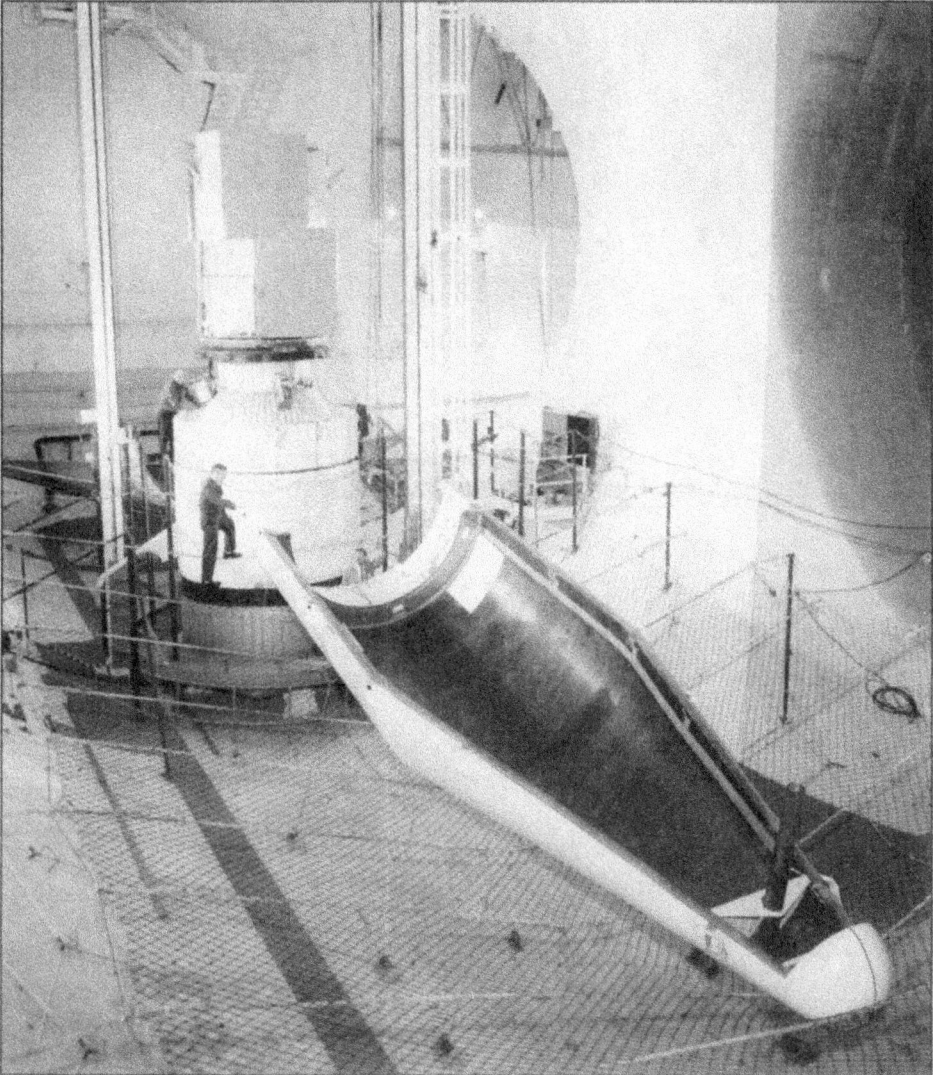

Image 246: OAO-2 shroud test setup in SPC No. 2. A steel base 10 feet in diameter was installed on the chamber floor. A metal shell spacer was attached to the base, and a mock Centaur forward bulkhead was attached to the spacer. The fairing and payload were then mounted to the forward bulkhead. Two 30- by 50-foot nylon nets were horizontally secured 11 feet above the chamber floor to catch the fairing halves after they were jettisoned.[527] (NASA C–1968–01258)

Image 247: Lewis researchers examine the base of the OAO-2 nose fairing after it was jettisoned in SPC No. 2. In the background is the transition adapter that was made especially for this heavy launch. It was placed between the Centaur and the payload so that the longer Agena shroud could be used on the Atlas/Centaur. (NASA C–1968–01259)

OAO-2 was launched into a geosynchronous orbit by AC-16 at an altitude of 480 miles on 7 December 1968. To obtain the desired thrust-to-weight ratio for lifting the heavy payload and shroud, the Atlas was secured to the launch pad for an extra 1.76 seconds. The launch went well, and the new fairing was jettisoned without problem.[528]

OAO-2 was a major accomplishment for the astronomy scientists, Goddard engineers, and the Centaur launch team. The satellite's 11 telescopes operated for over four years and provided a wealth of x-ray, ultraviolet, and infrared information on the stars.[529] After only a month, OAO-2 obtained over 20 times more ultraviolet data from stars than all of the sounding rocket studies of the previous 15 years combined.[530] It was also Centaur's largest payload to date. It required precise launch coordination, the use of the new transition adapter, and a new hybrid nose fairing. For their efforts the Lewis launch team received another NASA-wide Group Achievement Award.[531]

Image 248: Stamp issued by Sharjah (currently the United Arab Emirates) to celebrate the OAO-2 mission. It was the first successful space observatory and yielded a wealth of information on stars. (1964) (Don Hilger/Colorado State University)[532]

Image 249: Left to right: Abe Silverstein and Al Young were among seven Lewis staff members honored on 27 June 1969 for 40 years of federal service. Silverstein and Young had been the Altitude Wind Tunnel's (AWT's) original managers in the 1940s.[533] *The Apollo 11 crew would land on the Moon just a couple of weeks after the event, and both men would retire soon afterward. (NASA Lewis News, 3 July 1969)*

Following the OAO-2 tests in 1968, the SPC was relatively quiet for the next couple of years. The shop area was used to prepare the Pratt & Whitney TF-30 turbofan engine for testing in the Propulsion Systems Laboratory, but the test chambers remained unused. Lewis remained active, though, as the Apollo Program raced for the finish line. Centaur sent *Mariners 6* and *7* toward Mars, and Agena continued with the Space Electric Rocket Test II (SERT II) and several military launches. The center also returned to aeronautics and reinstated its airbreathing engines research program. Al Young and Abe Silverstein retired after 40 years of service. Silverstein said at the time, "As NASA engages in its second ten-year program, it may be important that the men whose decisions initiate the new long range projects be available to complete them. Since I do not think I can stretch my forty years of service into fifty, it is perhaps best for me and for the Lewis Center if I bow out now."[534]

Centaur Stumbles

At 10:40:05 p.m. on 30 November 1970, the AC-21 launch vehicle rose into the dark skies above Cape Kennedy carrying the follow-up to the successful OAO-2 space-based observatory. This satellite, OAO-B, included a Goddard-designed 36-inch telescope and spectrophotometer that was intended to measure ultraviolet energy emitted by stars from an orbit 466 miles above the Earth. The telescope was over twice the size of that on the OAO-2's and could observe stars eight times dimmer.[535]

When the command came to start the Centaur engines and eject the shroud, 1 of the 16 explosive nuts securing the fairing failed to release. During the 8 minutes and 20 seconds before the final jettison, the rocket was thrown off course by the extra weight of the 2,400-pound shroud. The rocket and satellite were destroyed when the Centaur plummeted back into the Earth's atmosphere.[536] It was the third official failure in Lewis's first 13 Centaur launches.

For the Centaur team, the loss was painful. Each OAO satellite was unique and required its own payload integration. Joe Nieberding explains, "They were not cookie cutters. They were different instruments, and there were folks who had worked on that one for seven years…and one bolt didn't open up on the shuttle, and it couldn't carry it all the way over. It was too heavy."[537]

A Lewis failure investigation team quickly went to work. The AC-25 launch was less than two months away. The Launch Vehicle Review Board, headed by H. Warren Plohr, quickly established that the mission failure resulted from the unsuccessful shroud jettison, but they could not determine the cause. Seventeen scenarios were analyzed during this initial study, but no single conclusive cause was found. The shroud system was the same as that used on the successful OAO-2 launch.[538]

Although not used to test the shroud before the AC-21 launch, the SPC was a key element of the failure investigation. The review board had recommended a redesign of the shroud, higher quality manufacturing and assembly, and intensified inspection methods.[539] In April 1971 William Prati, who ran the previous OAO shroud tests, oversaw investigations of a one-sixth-size shroud in SPC No. 2.

Image 250: A one-sixth-scale model of the failed AC-21 shroud is set up inside SPC No. 2. Investigators found that the spring actuators failed to jettison the shroud because of some restraint in the shroud. When the restraint was freed, it took over 8 minutes for the springs to release the shroud. The flight data on hinge strain and the accelerometer indicated that the latch system was the most probable cause.[540] (NASA C-1971-01491)

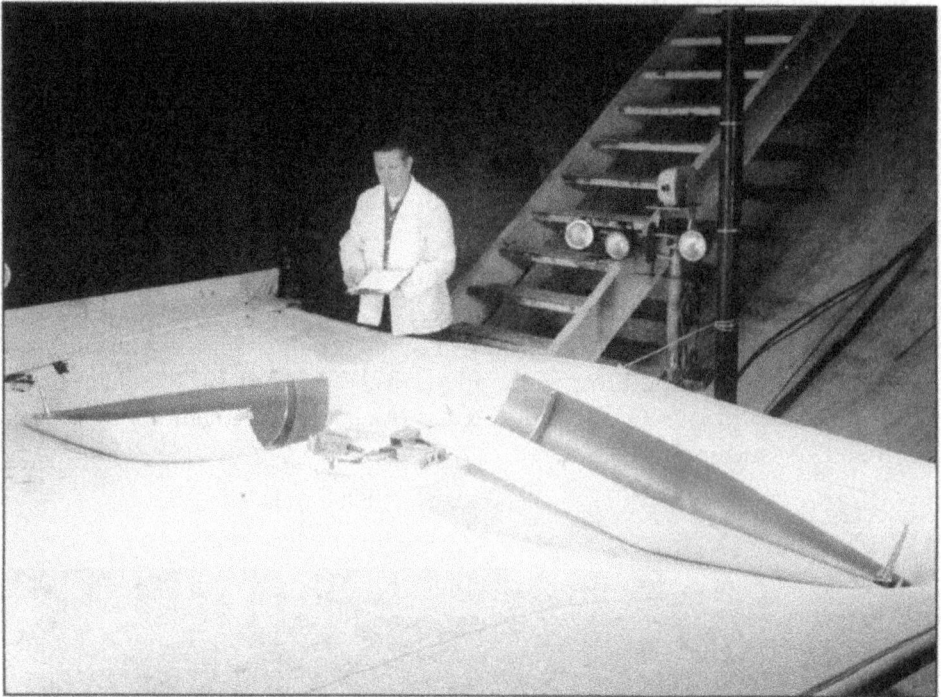

Image 251: The AC-21 shroud is investigated in SPC No. 2. The fairing was joined by 16 latches with no backup systems or redundancy. The board recommended adding mechanical redundancy to each nut, the self-containment of the separation nut and interface, additional inspections to ensure hardware conformity, and identification and removal of all nonflight hardware. Other recommendations were verification of cleaning fluids and the resolution of discrepancies of on-site fixes.[541] (NASA C–1971–01488)

Following the investigation, new explosive latches were designed for the fairing. Afterward, the entire full-scale separation system underwent extensive qualifications in SPC No. 2 in the spring of 1972. The runs were similar to the jettison tests prior to OAO-1 and OAO-2. The fourth and heaviest OAO satellite, renamed "Copernicus," was launched into orbit on 21 August 1972 by AC-22. Daniel Shramo, Centaur Program Manager, called it a "storybook flight."[542] The 4,900-pound Copernicus remained active for nine years and yielded even more spectral data than OAO-2.[543]

The launch was just weeks shy of the 10th anniversary of the program's transfer to Lewis. Afterward, Center Director Bruce Lundin said, "Heartbreaking failure and exhilarating success are merely the milestones in a long road of hard work, sacrifice, technical accomplishments, and comradeship that have marked this first decade of Centaur."[544]

Centaur Goes Modern

AC-22 was the final flight of the Centaur-D model spacecraft. Initially Centaurs had been made to order for the particular launch. In 1965 the Centaur-D became the first standardized model. Twenty of the Centaur-D's twenty-three flights were successful, but it was outdated by the late 1960s. Nieberding explained, "[The Centaur-D] was designed for *Surveyor*, for seven *Surveyors*. We got into those launches, and then the ATS came along, the Advanced Technology Satellites, AC-17 and 18, and the OAOs. Other users came along, but there was always like a three- or four-year future. It wasn't like a six-year, and then of course, the shuttle came along and then you only had a few more years, so there was never a long enough future to pump a lot of money into the vehicle. It was always, let's fix it for now and fix it for the next one."[545]

Centaur's successor was the D-1A. Its guidance system, telemetry, and electronics had been significantly upgraded and included a new 60-pound digital computer. To simplify the spacecraft and its interfaces, developers used an equipment module to store the payload.[546]

The Centaur fairing had been slightly modified during the upgrade to D-1A and had to be requalified in SPC No. 2 during winter 1972/73. Once again under the guidance of William Prati, the tests, which used mock-up *Mariner* and Intelsat IV payloads, verified the structural integrity and operation of the separation system.[547] In April 1973 the first launch of a D-1A sent the *Pioneer 11* spacecraft sailing passed Jupiter.

The new D-1A model could split its payload into two compartments—an experiment payload, which held the third-stage, and an equipment module, which stored support systems for the experiment. These were separated from the Centaur by an adapter. The system would be put to its limit during a series of High Energy Astrophysical Observatories (HEAOs) launches planned for the late 1970s.

Structural tests during the spring of 1975 in SPC No. 2 sought to determine the equipment module's flexibility, verify the load capacity, and purposely overload the structure to determine the point of failure. During this final phase of the test, the adapter failed before the equipment module. The tests continued with a new simulated HEAO adapter and successfully reached the module's failure point. The structural strength was deemed robust enough for the HEAO missions.[548] The three missions were highly successful and bridged the gap between the OAO telescopes of the 1960s and the Hubble Space Telescope launched in 1990.

Image 252: Centaur D-1A equipment module in SPC No. 2. The 19-foot-long HEAO satellites contained an experiment module on top and an equipment module attached to the bottom. The octagonal 3-foot-high, 7.5-foot-diameter equipment module, seen in the neck portion of this setup, included the avionics and other support systems for the HEAO experiments.[549] (NASA C–1975–01216)

Going Down Slow

The final HEAO test would be the last time that the SPC's test chambers would be used. The vacuum tank had been idle since 1967, and tests in SPC No. 2 had been staggered every few years. Two new facilities at NASA Lewis's auxiliary Plum Brook Station had come online in 1969. The Space Power Facility had become the world's largest vacuum chamber. It could handle the separation tests for the new, larger shrouds that were emerging at the time. The Space Propulsion Facility combined some of SPC No. 1's space simulation technology, such as the cold wall and radiant lamps, with the ability to fire large engines in a vacuum. SPC had served its purpose during the battle to get Centaur operational in the early 1960s and for the tricky shroud modifications for the larger payloads of the late 1960s. Although it operated through 1975, the facility's abilities had been superseded by these newer, more powerful facilities. The SPC would never be used again as a wind tunnel or a vacuum chamber.

Image 253: Space Propulsion Facility at Plum Brook Station. The facility, with its lid propped open to the left, combined some of SPC No. 1's space simulation methods and added the ability to fire large rocket engines in a vacuum. Its first tests in 1970 verified that Centaur could be flown without its boost pumps. In 1985 the Space Propulsion Facility was named a National Historic Landmark. (NASA C–1987–02664)

Endnotes for Chapter 11

501. Interview with Edgar M. Cortright, conducted by Rich Dinkel, 20 August 1998, Johnson Space Center Oral History Project, Houston, TX.
502. Cary Nettles to Virginia Dawson, "Centaur Notes," 8 May 2002, NASA Glenn History Collection, Cleveland, OH.
503. Interview with Seymour Himmel, conducted by Virginia Dawson, 1 March 2000, NASA Glenn History Collection, Cleveland, OH.
504. "Ranger B Preparations," *Lewis News* (17 July 1964), p. 1.
505. "Successes Three," *Lewis News* (28 February 1964), p. 3.
506. "3 Atlas-Agena Launches Assigned to Atlas-Centaur," *Lewis News* (6 January 1967), p. 1.
507. Walter Scott, "The Engineering Design of the Orbiting Astronomical Observatory" in *The Observatory Generation of Satellites* (Washington, DC: NASA SP–30, 1963).
508. Scott, "The Engineering Design of the Orbiting Astronomical Observatory."
509. "Large Load for Atlas-Agena," *Lewis News* (1 April 1966), p. 1.
510. "OAO Tests Here," *Lewis News* (3 September 1965), p. 4.
511. "First 100 Launches: Just the Beginning," *Lewis News* (15 March 1983), p. 4.
512. "Project OAO-B Press Kit," NASA press release No. 70-14, 29 October 1970, NASA Glenn History Collection, Cleveland, OH.
513. "Large Load for Atlas-Agena."
514. "Lewis Launches Wasp," *Lewis News* (24 June 1966), p. 8.
515. "Lewis Launches Wasp."
516. Courtney Brooks, James Grimwood, and Lloyd Swenson, *Chariots for Apollo: A History of Manned Lunar Spacecraft* (Washington, DC: NASA SP–4205, 1979), chap. 8.
517. Nettles to Dawson, "Centaur Notes."
518. "3 Atlas-Agena Launches," p. 3.
519. Charles Eastwood, Centaur—*Orbiting Astronomical Observatory Nose Fairing Altitude Jettison Test* (Washington, DC: NASA TM–X–2096, 1970), p. 2.
520. Lewis Research Center, *Atlas-Centaur AC–16 Flight Performance Evaluation for the Orbiting Astronomical Observatory OAO–II Mission* (Washington, DC: NASA TM–X–1989, 1968), p. 6.
521. "Centaur Is To Boost Biggest Payload Yet," *Lewis News* (8 November 1968).
522. "OAO Eyes Focus on Hidden Stars," *Lewis News* (20 December 1968), p. 3.
523. Warren H. Plohr, "Interim Investigations Conducted by the OAO–B Launch Vehicle Review Board," 12 January 1971, NASA Glenn Records, Box 46, Organizational Code 5800, H–7.
524. Larry Ross to Bob Arrighi, 9 June 2008, NASA Glenn History Collection, Cleveland, OH.
525. Plohr, "Interim Investigations," D–8.
526. Lewis Research Center, *Atlas-Centaur AC–16 Flight Performance Evaluation,* p. 107.
527. Eastwood, *Centaur—Orbiting Astronomical Observatory,* p. 8.

528. Lewis Research Center, *Atlas-Centaur AC–16 Flight Performance Evaluation,* p. 1.

529. Mark Wade, "OAO," *Encyclopedia Astronautica* (1995–2006) *http://www. astronautix.com/craft/oao.htm* (accessed 3 September 2009).

530. "OAO Meets Milestone," *Lewis News* (31 January 1969), p. 2.

531. "NASA Cites Atlas, Centaur for OAO II," *Lewis News* (6 June 1969).

532. Hillger, Don, "Orbiting Astronomical Observatory Satellites" (NOAA/ NESDIS/STAR/RAMMB, CIRA, 2004–2008), *http://www.cira.colostate.edu/cira/ RAMM//hillger/OAO.htm* (accessed 7 October 2009).

533. "Director Marks Forty Years," *Lewis News* (3 July 1969), p. 1.

534. "Dr. Silverstein Retiring," NASA Lewis press release 69–61, 22 October 1969, NASA Glenn History Collection, Cleveland, OH.

535. "OAO To Study Stars," *Lewis News* (23 October 1970).

536. "OAO Failure," *Lewis News* (18 December 1970), p. 3.

537. Interview with Joe Nieberding and Robert Gray, conducted by Virginia Dawson, 9 November 1999, NASA Glenn History Collection, Cleveland, OH.

538. Plohr, "Interim Investigations," H–18.

539. John E. Naugle to Deputy Administrator, "OAO–B Launch Vehicle Failure Review," 17 June 1971, NASA Glenn History Collection, Cleveland, OH.

540. Plohr, "Interim Investigations."

541. Plohr, "Interim Investigations."

542. "1962–1972 The First Decade of Centaur," *Lewis News* (20 October 1972).

543. National Space Science Data Center, "OAO–3," NSSDC ID: 1972–065A, *http://nssdc.gsfc.nasa.gov/nmc/spacecraftDisplay.do?id=1972-065A* (accessed 7 April 2008).

544. Bruce Lundin, "Director Praises Centaur Program," *Lewis News* (20 October 1972).

545. Nieberding and Gray interview, conducted by Dawson, 9 November 1999.

546. "Single Spacecraft To Visit Two Planets," *Lewis News* (2 November 1972), p. 2.

547. William M. Prati, *Centaur D–1A Nose Fairing Jettison Test* (Washington, DC: NASA TM X–73450, 1976), p. 1.

548. Thomas Niezgoda, *D–1A Equipment Module Structure Test* (Washington, DC: NASA TM X–73401, 1976).

549. Wallace Tucker, The Star Splitters: *The High Energy Astronomy Observatories* (Washington, DC: NASA SP–466, 1984).

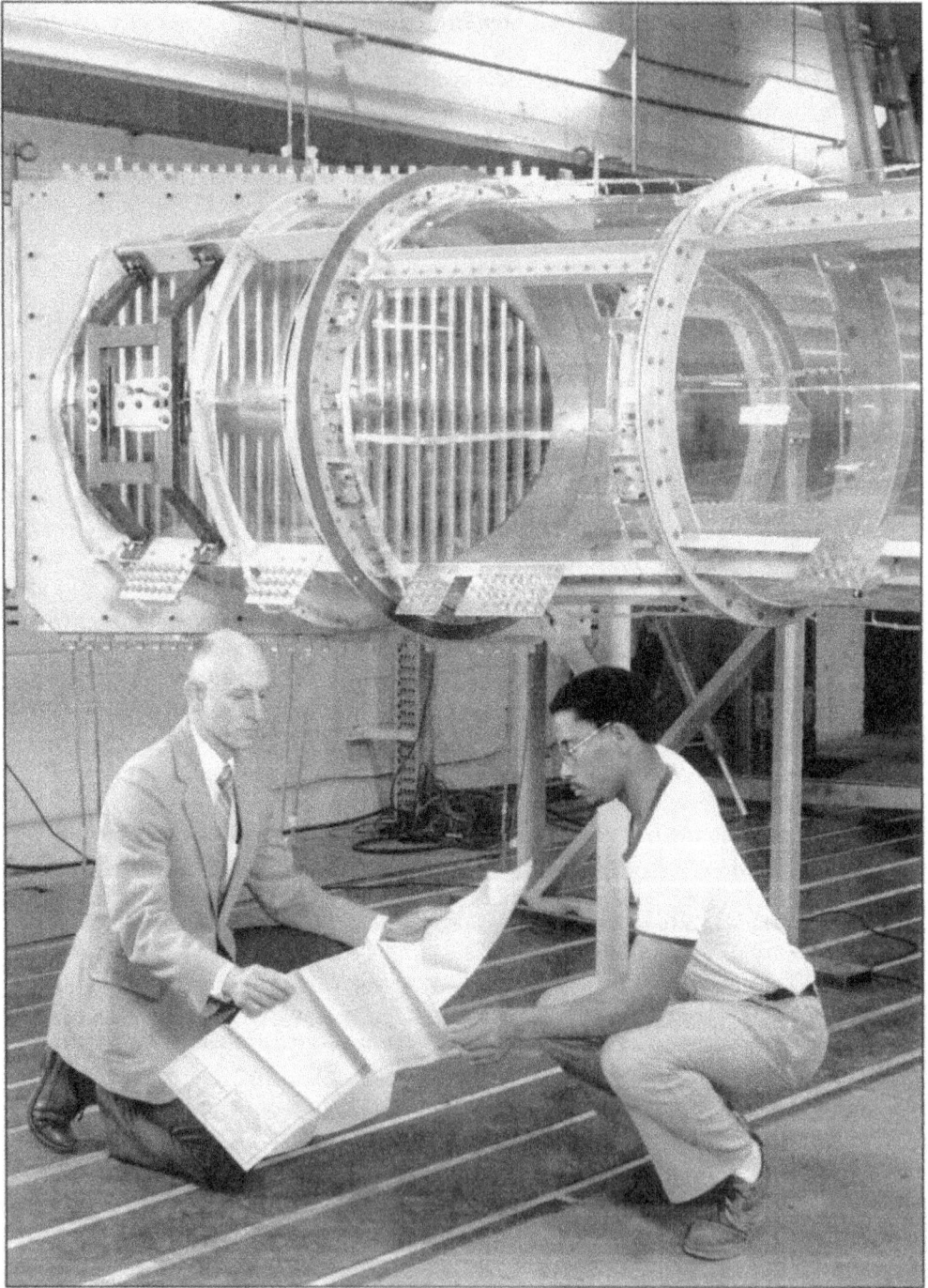

Image 254: Researchers examine blueprints for a model of a new Altitude Wind Tunnel. The center proposed to convert the Space Power Chambers, which had not been used in several years, back into a wind tunnel configuration. (NASA C–1985–03172)

Chapter 12

Where the Wave Finally Broke | The Idle Years (1975–2005)

"We're living in difficult times," Bill Harrison told the 1973 graduating apprentices, "and the measure of our success in the future will lie in our ability to accept change and bad times and emerge stronger than before." The December 1972 Apollo 17 mission brought a close to NASA's halcyon days of the 1960s. Within days of the splashdown, large space programs were being cancelled across the Agency. The effect was felt immediately by the staff at the NASA Lewis Research Center. The closure of Plum Brook Station with its two new world-class facilities was announced on 6 January 1973. Over the next two years, over 750 civil servants were either let go or retired.[550] Many of the graduating apprentices had already been informed that the center could not hire them because of the budget cuts. Harrison lamented, "Certainly in today's world there is no lack of challenge, only the funding to pursue each and every problem."[551]

The center did not have a clear mission at the time, and its role in the development of the space shuttle would be minimal. The state of the center continued to deteriorate in the late 1970s with declining budgets and staffing. The steady reduction of the workforce and the looming threat of center closure resulted in the nadir of employee morale. One area that did not suffer was the Launch Vehicles Division. It continued to send Centaur rockets into space carrying interplanetary payloads such as the *Pioneer* and *Voyager* spacecraft.

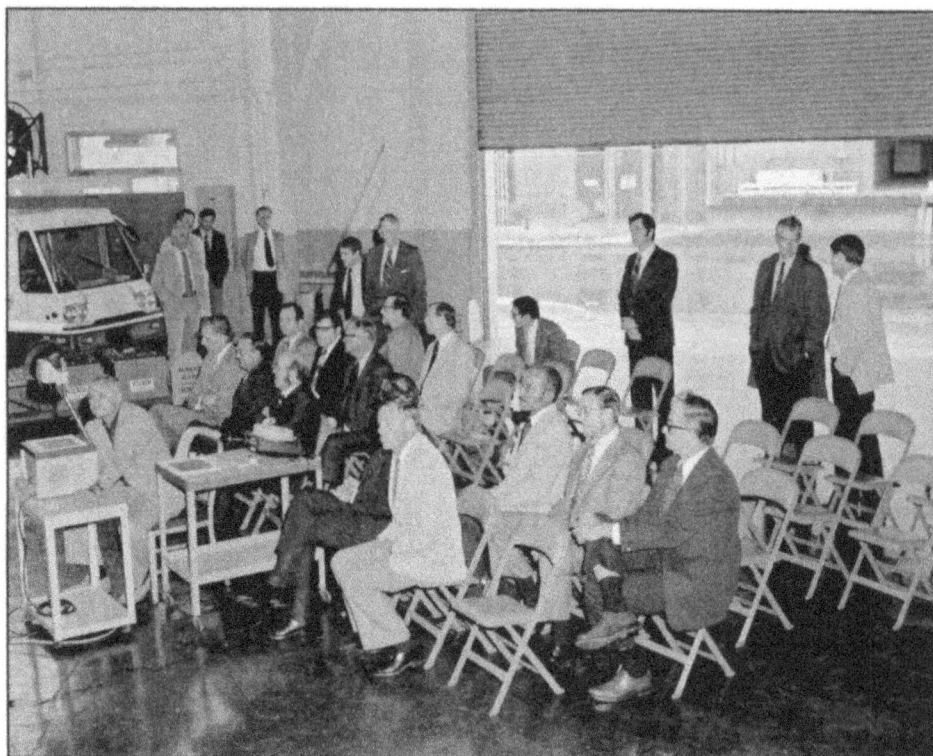

Image 255: The 1970s were not kind to NASA Lewis. Without a clear mission or stable budget, the center sought nontraditional research projects such as wind energy, electric automobiles, and quiet jet engines. In this photograph, several NASA center directors and NASA Head-quarters managers gather in the SPC shop area to listen to a description of the Electric Vehicle Project at Lewis. (NASA C–1976–04044)

Bruce Lundin, who had succeeded Abe Silverstein as Center Director, sought out renewable energy and aeronautics research projects. These included wind turbines, solar cells, Stirling engines, and electric automobiles. The Space Power Chambers (SPC) shop area was used as the base for the electric vehicle studies. As early as 1979, Lewis began long-term planning to combat NASA's downturn. Despite extraordinary efforts, many of the proposed new programs could not be undertaken because of budget constraints.[552] One of those failed enterprises was the proposed reinstitution of the SPC in its original Altitude Wind Tunnel (AWT) configuration.

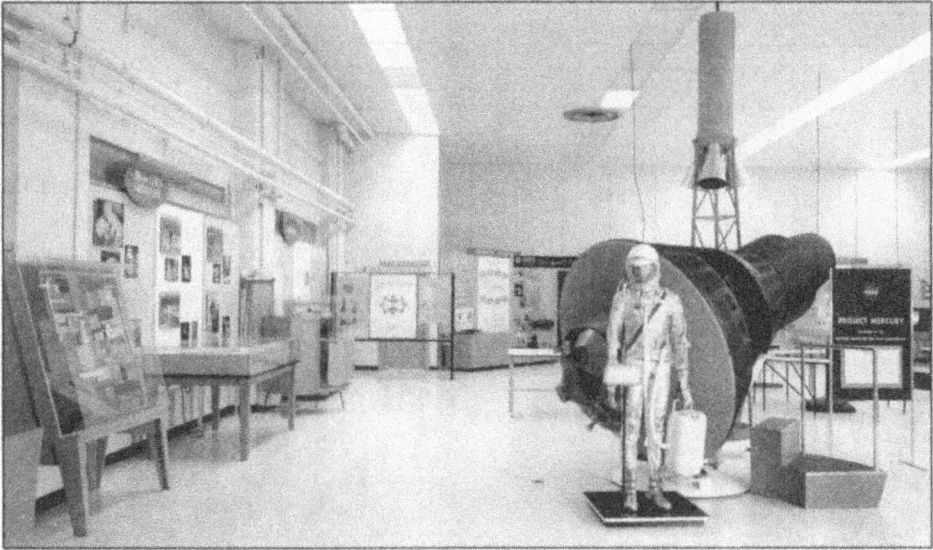

Image 256: Part of the AWT Exhauster Building became a clean room known as the Solar Power Laboratory in the mid-1960s. In July 1970, however, the structure opened as the Aerospace Information Display building. It contained models, hardware, and exhibits that had been used at the space science fairs of the early 1960s. (NASA C–1970–02258)

Image 257: In 1975 the Aerospace Information Display was expanded and renamed the "Visitor Information Center." The exhauster equipment had been shipped to NASA Marshall. A large lobby was created, and the annex became an assembly room. NASA retiree Calvin Weiss estimated that annual visitors increased from 1,000 to 50,000 during that first year.[553] (NASA C–1978–1648)

Image 258: The AWT shop area was used to store and work on vehicles for the NASA and Energy Research and Development Administration Electric Vehicle Project. (NASA C–1977–01813)

Image 259: NASA researchers in front of a van used to support the Electric Vehicle Project Office. The van was outfitted with battery chargers and other specialized test equipment. (NASA C–1977–01797)

Space Power Chambers Becomes a Garage

Through an interagency agreement between NASA and the Energy Research and Development Administration (ERDA),[||] Lewis engaged in several energy-related programs in the mid-1970s, including the Electric Vehicle Project. Lewis investigated new drivetrains, improved powertrains, and enhanced the overall engine. As part of the project, Lewis also tested a fleet composed of every commercially available electric car. The SPC shop area was used to store and work on these vehicles.[554]

Battery longevity was the crucial element for the electric automobile's range. Lewis applied its experience with the long-life batteries used for satellites to these new cars. The result was a nickel-zinc battery that offered almost twice the range of traditional automobile batteries. The range was demonstrated during test runs in a utility van using both types of batteries.[555] Like many of the center's energy research programs of the 1970s, the concept was proven but never taken on by industry. A combination of factors, including consumer reluctance and the eventual leveling of gas prices, resulted in these programs being shelved indefinitely. These efficiency programs are occasionally revisited when oil prices rise precipitously.

A New Propulsion Wind Tunnel

In 1982 former Lewis propellant researcher and Titan/Centaur manager Andy Stofan returned from an assignment at NASA Headquarters to assume the position of Center Director of the NASA Lewis Research Center. Stofan assembled a team to create and implement the center's first strategic plan. Lewis set its sights on five new major programs—the Space Station Freedom Electrical Power System, a Centaur to be carried on the space shuttle, the Advanced Communications Satellite, the Advanced Turboprop, and a reinstitution of the AWT.[556]

Lewis advocates for the AWT claimed at the time that "there are no propulsion wind tunnels in the free world that provide a subsonic, standard operating envelope with true altitude pressure and temperature." The proposed reinstitution of the AWT would increase the tunnel's speed to Mach 0.9 at an altitude of 55,000 feet and its temperature to −20°F for jet engine testing, icing research, and noise-suppression studies.[557]

[||]ERDA was later folded into the Department of Energy in much the same way that the NACA was into NASA.

In the early 1970s Bruce Lundin had called upon trusted Lewis veterans to put together a cost study for the reinstitution of the AWT for vertical or short takeoff and landing (V/STOL) testing. Harold Friedman, Milt Beheim, J. C. Lovell, W. E. Emley, and Robert Godman authored a report that recommended expanding the diameter of the test section to 24 feet and the length to 30 feet. The report, which converted the costs of the original components into 1976 figures, estimated that the reinstitution of the tunnel would cost $39 million.[558] This proposal was not acted upon.

The second look in the 1980s into reinstituting the AWT, however, was much more ambitious. New and envisioned future aeronautics technologies such as V/STOL, high-speed turboprops, highly survivable military aircraft, and high-speed rotorcraft could be studied. These new aircraft systems would require a wind tunnel that could support large-scale test articles and propulsion systems. High subsonic speeds, altitude conditions, and simulated inclement weather were necessities.[559] Lewis's dormant AWT/SPC seemed like an excellent location for this new facility.

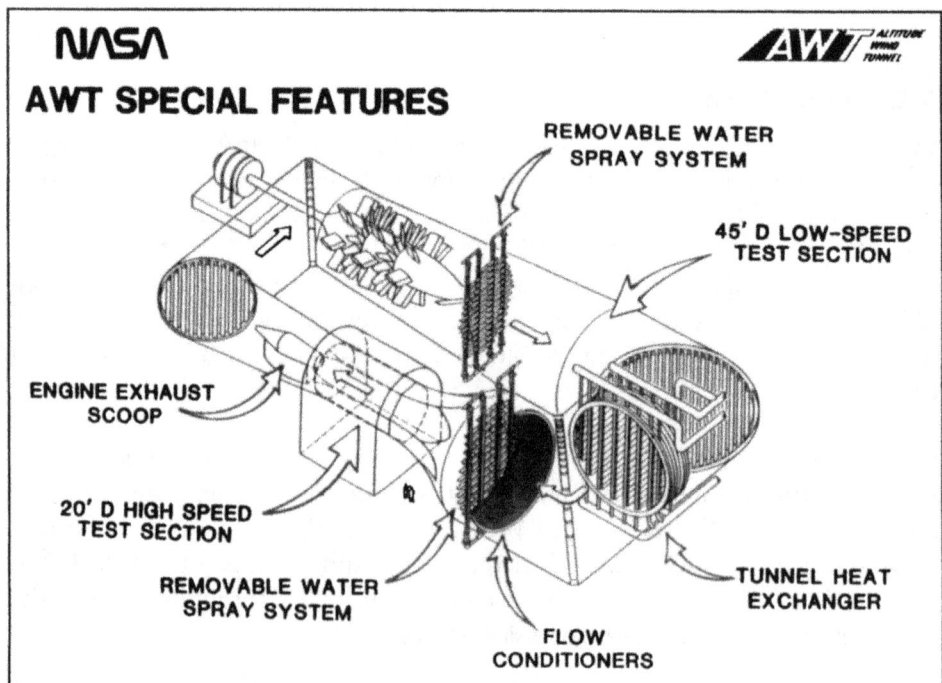

Image 260: Special features of the proposed new AWT included low- and high-speed test sections, an exhaust scoop, cooling coils, and two spray-bar systems. This is one of many drawings created by Sverdrup Corporation for their extensive Preliminary Engineering Report. (1984) (NASA Glenn)

Image 261: Members of the AWT Project Office inside the tunnel. The office consisted of several groups. The Engineering Office executed Construction of Facilities tasks such as the Preliminary Engineering Report, final design, studies, and construction management. The Research Office ensured that the facility met all the proposed specifications and requirements. The Systems Office guaranteed that the facility was safe and efficient.[560] *(NASA C–1984–00731)*

The AWT Project Office, led by Roger Chamberlin, was established in 1980 to oversee the project. The office drew deeply from Lewis's extensive experience with propulsion tunnels as well as from the tunnel expertise of the NASA Langley and NASA Ames research centers, Boeing, and the Arnold Engineering Development Center. Different teams addressed the requirements for aerothermodynamics, the creation of icing conditions, the establishment of a dynamic math model, and the acoustical problems.[561]

The center contracted with Sverdrup Corporation in 1981 to explore the options and expenses for reinstituting the AWT. By November 1984 Sverdrup had completed the Preliminary Engineering Report. The 15-volume document was the most comprehensive engineering report ever created for a NASA facility.[562] It analyzed all of the tunnel's components and explored various options for the modifications. The report affirmed Lewis's hopes. The existing infrastructure was deemed robust enough to be the basis for the new tunnel.[563]

Unlike the original tunnel design, the new tunnel would allow various inclement weather testing and acoustical measurement instruments. Lewis's icing program was so successful in the 1940s and 1950s that it had been retired in the early 1960s. The evolution of aircraft wing design in the 1970s resulted in calls for new icing studies. In the 1980s, the program was resurrected. The Icing Research Tunnel was brought back online and maintained a busy schedule. It was felt that the new AWT could alleviate some of the Icing Research Tunnel backlog and permit the testing of larger components and altitude simulation.

The AWT's new test section would have a slotted throat and honeycomb screen upstream to smooth the airflow.[564] Since the tunnel's internal elements had been removed during the creation of the SPC in 1962, a new test section, exhaust scoop, heat exchanger, two-stage fan system, and turning vanes would have to be installed. In addition, the steel bulkheads used to create the vacuum chamber would have to be removed.[565]

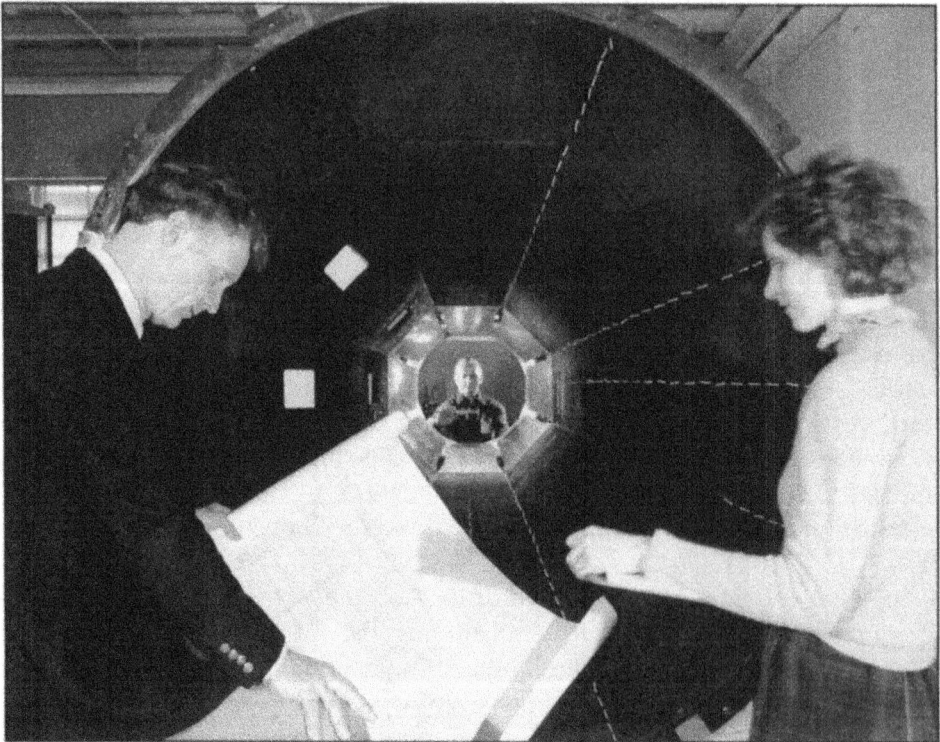

Image 262: Members of the AWT Project Office examine drawings for the proposed reinstitution of the AWT. A researcher is seen through the contraction in the high-speed leg model. (NASA C–1985–01572)

Image 263: AWT model assembly in cell CW-22 of the Engine Research Building. The model included the high-speed test section near the middle and a corner with a turning vane at the right. (NASA C–1985–02361)

The tunnel's drive motor, refrigeration system, shell, and auxiliary buildings all remained. These items were valued at $30 million in the 1980s. Installation of the new components including spray bars was estimated at $20 million. The AWT Project Office claimed, "For a $20 million investment, a $50 million facility could be obtained which will contribute significantly to many NASA programs in regions of the flight envelope not currently available in other U.S. facilities."[566]

In mid-March of 1984, a seminar on wind tunnel modeling was held at NASA Lewis. As part of the planning process a 1/10th scale model of the tunnel had been built to test various components and systems. Members of the AWT Project Office discussed the AWT reinstitution and modeling status. Representatives from NASA's aeronautics centers, Langley and Ames, and manufacturers Calspan Corporation and Boeing related their respective modeling programs. This was followed by detailed discussions of the AWT's drive system, acoustics, controls, and icing simulation.[567]

Image 264: A Centaur 6A rocket was placed inside SPC No. 1 in 1963 for long-duration space environment testing. Although the studies ended in the late 1960s, the rocket remained inside the chamber until August 1984. It was decided to remove the Centaur when the center considered returning the facility to its original wind tunnel configuration. (NASA C–1984–04470)

All for Nothing

H. Harvey Album, Chairman of the Congressional Advisory Committee on Aeronautics Assessment, formed an Ad Hoc Altitude Wind Tunnel Advisory Subcommittee to review the proposed AWT reinstitution project in depth. The eight-member subcommittee was composed of experts on large transport aircraft, rotorcraft, propulsion systems, and military aircraft.[568]

Following a report by the AWT ad hoc subcommittee, the Congressional Advisory Committee on Aeronautics Assessment decided to cancel fiscal year 1986 funding for the reinstitution studies. The AWT Project had consumed a substantial amount of personnel and financial resources. An estimated $5 million and 100 man-years were scheduled for the upcoming fiscal year, and it appeared that the actual reinstitution of the tunnel would exceed the $160 million already proposed. In addition, the subcommittee calculated that the tunnel would cost over $5 million per year just to operate.[569]

The cancellation was also based on technical reasons. The Congressional Advisory Committee on Aeronautics Assessment argued that the AWT would duplicate the capabilities of the Arnold Engineering Development Center and other tunnels. The committee also questioned the AWT's potential. They considered the 20-foot-diameter test section to be too small to test helicopter rotors, propfans, or large propulsion systems. They felt that it would be capable of inlet propulsion integration studies only on small engines. Smaller size models, aircraft, engines, rotors, and propulsion systems could be tested in other facilities or during test flights.[570]

The committee felt that full-scale models would produce questionable results by not allowing enough open space for the proper amount of airflow through the AWT test section. They also felt that the test section would only be capable of testing smaller models or engines because of this blockage. Propulsion systems can really only be tested with full-size engines. They concluded that the range of altitude and Mach numbers in the AWT spectrum did not allow Mach 0.9 at an altitude of 25,000 feet and that its Mach number range was limited at altitudes of 30,000 feet. Finally, the committee felt that the AWT's proposed inclement weather testing would only work on smaller engines.[571]

The committee suggested that the Arnold Engineering Development Center's 16T, 16S, and Aeropropulsion Systems Test Facility could adequately test large-scale engine systems. They, along with NASA Ames's 11-foot and 14-foot tunnels, would be high Reynolds number facilities. The Aeropropulsion Systems Test Facility could be used for high-altitude,

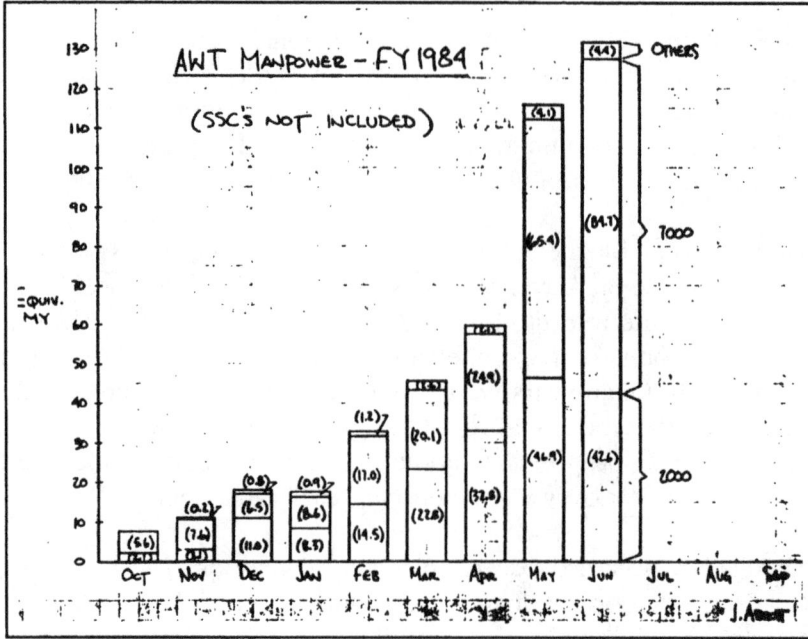

Image 265: Increase in manpower needed for the AWT reinstitution project. (NASA Glenn)

low-Mach-number fighter jet testing; it would be suitable for engine inlet tests in its freejet mode, and general engine tests could be done in its direct-connect section. Lewis's Icing Research Tunnel was seen as a sufficient alternative for full-scale wing section and inlet testing, although it was acknowledged that the AWT would be superior for propulsion testing.[572]

The AWT Project Office prepared a three-page response to the Congressional Advisory Committee on Aeronautics Assessment's findings. They addressed the fact that full-scale engines or models would reduce the Reynolds number significantly. The AWT group pointed out that aerodynamic tunnels and propulsion tunnels had different test requirements. Aerodynamic testing required having Mach and Reynolds numbers as close as possible to actual aircraft. This was best accomplished with smaller models in either atmospheric or pressurized tunnels. Propulsion tests, however, required the matching of altitude and pressure in addition to Mach number and Reynolds number. Propulsion tests also permitted higher blockages, so larger or full-scale models could be used. This meant that large or full-scale Reynolds numbers could be achieved at simulated altitudes.[573]

John Murphy, the Office of Aeronautics and Space Technology's Assistant Administrator of Legislative Affairs, refuted the committee's claims in a letter to the Chairman of the U.S. House Subcommittee on Transportation, Aviation, and Materials. He claimed that the AWT's 20-foot test section would allow testing of full-scale turboprops and a half-span of an F-15 with its F-100 engine. Neither of these could be accomplished in any other tunnel in "the free world."[574] Nonetheless, the decision stood, and the AWT was left fallow.

Antenna Field

The tunnel would remain dormant, but one of Andy Stofan's other major programs from the 1980 master plan, the Advanced Communications Technology Satellite, resulted in some activity for the facility. In 1982 the Shop and Office Building was converted into an antenna testing facility and was renamed the "Microwave Systems Laboratory." The new laboratory

Image 266: This Near-Field Antenna Test Facility in the high bay was used to test new sophisticated higher frequency space communications antennas and proof-of-concept antennas for the Advanced Communications Technology Satellite.[575] The foam pyramids along the walls absorbed microwaves. (NASA C–1984–00747)

Image 267: Bill Darby, a mechanical engineering technician, stands beneath an inflatable membrane antenna inside the Near-Field Antenna Test Facility in the high bay. The 4- by 6-meter offset-parabolic inflatable membrane reflector and software were developed to demonstrate that a novel ground station composed of an array of relatively small apertures could economically replace a single, expensive tracking ground station. (NASA C–2004–01883)

311 Where the Wave Finally Broke

consisted of the Near-Field Antenna Test Facility in the high bay and the Far-Field Antenna Test Facility in the former SPC No. 1 control room. The walls of the high bay were covered with row after row of foam pyramids to absorb any microwave rays that escaped the antenna.

In this setting researchers could scan a 22- by 22-foot area just a few thousandths of an inch away from the surface. The ability to study these large antennas at such a close distance allowed the researchers to extrapolate the data to that of an antenna beam's behavior when connecting to orbiting communications satellites. The only other alternative for this testing would require miles of distance between the antenna and probe in difficult test conditions.[576] The facility was expanded in 1991 and remains active today. The tunnel, however, remained silent and empty.

Endnotes for Chapter 12

550. Virginia P. Dawson, *Engines and Innovation: Lewis Laboratory and American Propulsion Technology* (Washington, DC: NASA SP–4306, 1991), p. 262.
551. Bill Harrison, "Address to Apprentice School Graduation," 8 June 1973, NASA Glenn History Collection, Cleveland, OH.
552. "NASA Five Year Planning (FY–1980–84)," January 1979, NASA Glenn History Collection, Cleveland, OH.
553. Calvin Weiss, discussion with Bob Arrighi, NASA Glenn Research Center, 24 October 2008.
554. "The ERDA–NASA Energy Activities Team," *Lewis News* (14 May 1976), p. 3.
555. "New Technology for Transportation," *NASA Spinoff 1977* (1977), p. 90.
556. Dawson, *Engines and Innovation,* p. 214.
557. Donald Boldman, Royce Moore, and Rickey Shyne, *Experimental Evaluation of Two Turning Vane Designs for Fan Drive Corner of 0.1 Scale Model of NASA Lewis's Proposed Altitude Wind Tunnel* (Washington, DC: NASA TP–2646, 1987), p. 2.
558. Harold Friedman, "Altitude Tunnel for STOL Engine Testing," memo for the record, 15 July 1971, NASA Glenn History Collection, Cleveland, OH.
559. "AWT Project Management Report: July 1985," 1 August 1985, NASA Glenn History Collection, Cleveland, OH. p. 2
560. "AWT Project Management Report: July 1985."
561. John Abbott, et al., *Analytical and Physical Modeling Program for the NASA Lewis Research Center's Altitude Wind Tunnel* (Reston, VA, AIAA–85–0379, 1985), p. 3.
562. "AWT Narrative," c1985, NASA Glenn History Collection, Cleveland, OH.
563. "Engineering Services" *Lewis News* (31 December 1981).
564. Abbott, et al., *Analytical and Physical Modeling Program.*
565. Boldman et al., *Experimental Evaluation.*
566. "Altitude Wind Tunnel Rehabilitation," NASA Glenn History Collection, Cleveland, OH.
567. *Wind Tunnel Modeling Seminar* (Washington, DC: NASA TM–104978, 1984).
568. Ad Hoc Altitude Wind Tunnel Advisory Subcommittee, "Assessment of the Altitude Wind Tunnel at NASA Lewis Research Center," 25 February 1985, NASA Glenn History Collection, Cleveland, OH.
569. Congressional Advisory Committee on Aeronautics Assessment, "NASA Aeronautics Budget for FY86," NASA Glenn History Collection, Cleveland, OH.
570. Congressional Advisory Committee on Aeronautics Assessment, p. 1.
571. Congressional Advisory Committee on Aeronautics Assessment, p. 2.
572. Congressional Advisory Committee on Aeronautics Assessment, p. 4.
573. AWT Subcommittee, "Response to CACA on Their Comments Relative to 'Alternative Facilities'," NASA Glenn History Collection, Cleveland, OH.

574. John F. Murphy, Assistant Administrator for Legislative Affairs, to Honorable Dan R. Glickman, Chairman of the Subcommittee on Transportation, Aviation, and Materials, March 1985, NASA Glenn History Collection, Cleveland, OH.

575. Larry Ross, "Year in Review—Space," *Lewis News* (December 1982).

576. "New Test Facility Investigates Large Space Antennas," *Lewis News* (16 December 1983).

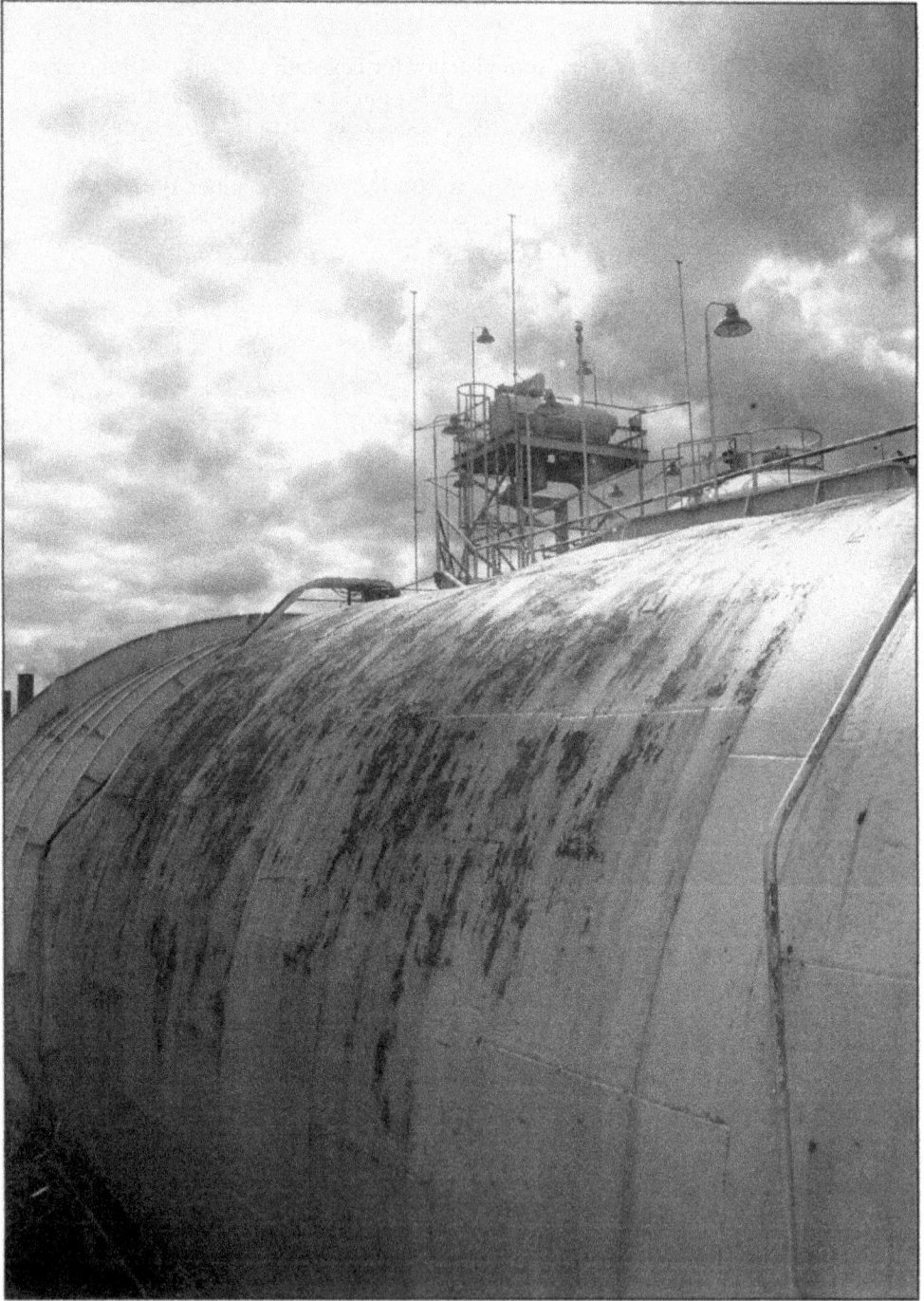

Image 268: The Altitude Wind Tunnel/Space Power Chambers as it appeared in August 2005. The facility began suffering from neglect in the 1990s. In 2004, the NASA Glenn Research Center began taking steps to demolish the facility. (NASA C–2007–02571)

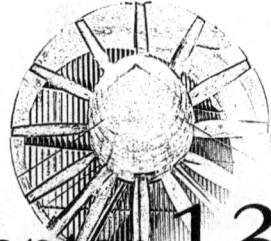

Chapter 13

Death Knell | The Final Days of the Altitude Wind Tunnel/Space Power Chambers (2005–2009)

After the Altitude Wind Tunnel (AWT)/Space Power Chambers (SPC) facility spent almost its entire lifetime evolving to stay current, the 1985 cancellation of the proposed restoration of the tunnel portended the fate of the facility. Maintenance of the facility seems to have ceased in the early 1990s. The exterior shell began rusting, and birds started using the SPC dome for shelter. The test chamber room became littered with old equipment. The original control room was steadily cannibalized, and the space was used as a storage room. The tunnel's primary asset, its 1-inch-thick inner steel shell, however, remained in fairly good shape.

In 2003 NASA began examining its assets and infrastructure. The Agency was about to embark on its mission to return to the Moon. The examination had two facets: a look at NASA's ability to meet the nation's aerospace testing needs and a determination of which unused facilities and structures could be demolished. Since the AWT was not being used, it was omitted from the former focus but was a prime candidate for the latter. In this atmosphere, it was not surprising when NASA Glenn** decided to demolish the AWT/SPC. The center had no plans to use the AWT/SPC's footprint, but an internal study indicated that upkeep costs such as repainting the exterior would rival those of complete demolition. The days of the once-essential AWT/SPC were officially numbered.

**The name of the center was changed from the NASA Lewis Research Center to the NASA John H. Glenn Research Center on 1 March 1999.

Image 269: Glenn Chief Architect Joe Morris leads a tour of the AWT/SPC in August 2005. The tunnel had been idle for 30 years, and its vast interior was used to store such items as a mirror ball, Christmas Club decorations, and a Star Wars R2D2 replica. (NASA G6QH9170)

Wind Tunnels Become Endangered

The construction of wind tunnels at NASA had been declining for many years. The Unitary Plan Act of 1949 sought to build complementary tunnels at NACA, military, industry, and university sites. Following this wave of Unitary Plan Act tunnels in the mid-1950s, which included the 10- by 10-Foot Supersonic Wind Tunnel at Glenn, NASA's construction of new large wind tunnel's tapered off. Its most recent major tunnel was the National Transonic Facility, which began operating at the NASA Langley Research Center in 1983. Glenn's last wind tunnel was the Hypersonic Test Facility, which became active at its Plum Brook Station in 1971.

In 2003 the Rand Corporation was hired to analyze NASA's ability to serve national research needs. The study focused on the Agency's wind tunnels and propulsion test facilities. It identified 29 of the 31 active NASA tunnels as unique, and it recommended that they be maintained. Although the capabilities of some of these facilities were duplicated by other national and international facilities, the Rand paper claimed that each tunnel was so specialized that schedule backlogs were common.[577] Since the AWT was not in service at the time, it was not included in the Rand study.

In April 2004 President George W. Bush announced his Vision for Space Exploration, which would send astronauts first to the Moon and then to Mars. Aeronautics programs at Glenn and throughout NASA were slashed. Glenn endured a tenuous period of adjustment. There were employee buyouts, few new hires, budget cutbacks, a massive reorganization, and rumors that the center would close.

In March 2005 the U.S. House Subcommittee on Space and Aeronautics, led by Congressman Ken Calvert of California, held hearings on the future of aeronautics at NASA. The hearings were called to address the drastic cutbacks in the fiscal year 2006 aeronautics budget, but they also examined longer-term decisions such as the future of civil aeronautics research and development, the effect of workforce reductions on the Agency's aeronautics capabilities, and the preservation of wind tunnel and propulsion test facilities. Among the subcommittee's recommendations was that NASA continue to dispose of underutilized facilities. "Some of these facilities are not unique, and long-term fixed costs could be reduced through consolidation and deactivation."[578] This was almost in direct contrast to the Rand study two years before.

The Die Is Cast

It was in this atmosphere that the decision was made to remove the AWT/SPC. In 2004, for the first time in its history, NASA Headquarters allocated funds for the demolition of unused facilities and asked its centers to submit projects for consideration. Glenn proposed the removal of nine buildings, including the tunnel portion of the AWT/SPC. No testing had been conducted inside the facility since the mid-1970s. Although the AWT/SPC was unique on the basis of size alone, Glenn felt that the $93,000 annual maintenance costs for the facility were exorbitant.[579] A 2004 estimate for exterior repairs and repainting of the AWT/SPC were over $4.5 million, and the tunnel would have to be repainted every 8 to 10 years.

Glenn considered alternatives other than demolition of the facility, but relocation of the facility was unrealistic, and there had been no interest expressed by other government agencies or private companies in the use of the facility since the mid-1980s. There would not be much need for a new wind tunnel with NASA's new Vision for Space Exploration agenda. Rehabilitation of the tunnel for NASA's use was considered to be a "waste of millions of dollars" since the facility was "no longer needed."[580]

NASA Headquarters concurred with Glenn's decision and advocated the proposed demolition.[581] Glenn spent the next two years working out the demolition plans and soliciting bids from construction companies. Design services were obtained, and demolition plans were created. Bids to perform the work were solicited, and the $3.5 million contract was awarded to the Pinnacle Construction Development Group in 2007. Pinnacle contracted with Brandenburg Industrial Services from Chicago to perform the actual demolition.[582]

Image 270: Throat section of the AWT as seen on 5 April 2005 during a photographic survey prior to the tunnel's demolition. The stairs and telephone were installed for the Project Mercury tests in the late 1950s, including the Multi-Axis Space Test Inertia Facility (MASTIF). A bulkhead sealed the leg off from the former test section on the other side. (NASA C–2007–00377)

Image 271: Exterior of SPC No. 1 in 2005. The inner tunnel shell was in relatively good shape, but the exterior weather shell, seen to the left, was heavily rusted. (NASA C–2005–01670)

Image 272: Interior of SPC No. 1. Over the years, rain entered the open portals in the dome and collected on the chamber floor. This resulted in heavy rust damage along the bottom of the otherwise robust chamber. (NASA C–2005–01646)

Image 273: This panel and the acoustical tile walls are the only remaining signs of the former AWT control room. The room was used to operate SPC No. 2 tests in the 1960s and 1970s, but the control panels and measurement equipment were removed over the years. The space is now used as a storage room. (NASA C-2007-00398)

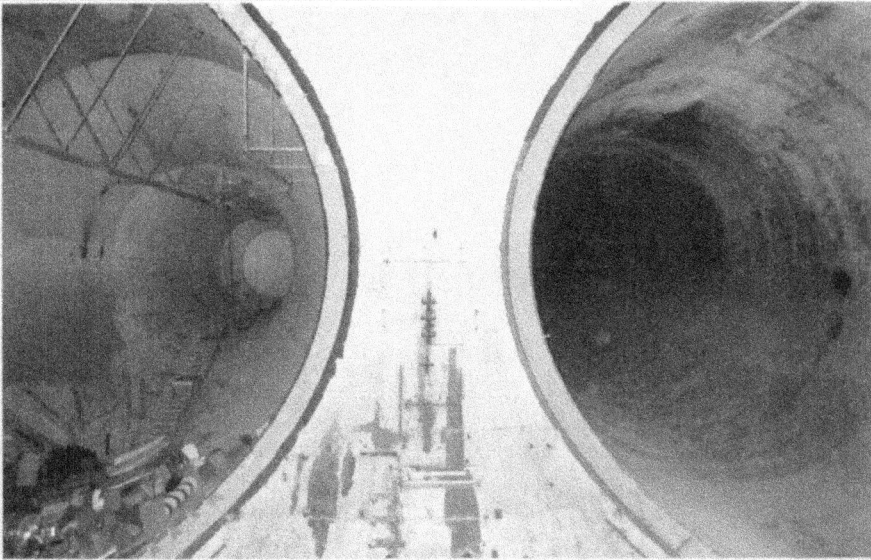

Image 274: Interior of the AWT/SPC as it appeared in 2005. The section to the left, where the MASTIF had been set up 45 years before, was later used to store holiday decorations (viewed from the west). (NASA C–2005–01615)

Image 275: A section of the inner shell was cut away in 2005 to view the steel mesh and fiberglass insulation. The ties that affixed the insulation to the mesh were made from an asbestos-based material. Left in place, the ties posed little danger, but extra safety precautions were used when dismantling the tunnel.[588] (NASA C–2005–01625)

Image 276: View into the former test section. The area had been used to store large antenna equipment in recent years. (NASA C–2009–00763)

Image 277: This area immediately beyond the test section had become a storage area in recent years. The door at the far end opened into SPC No. 1. (NASA C–2007–02565)

NASA's Historic Properties

The 1966 National Historic Preservation Act created the National Register of Historic Places to list buildings, structures, and sites worthy of preservation because of their contributions to U.S. history. NASA currently has 20 National Historic Landmarks and many pieces of property on the National Register of Historic Places. NASA Glenn has two current National Historic Landmarks, the Zero Gravity Facility at Lewis Field and the Space Propulsion Facility at Plum Brook Station. A third, the Rocket Engine Test Facility, was demolished in 2003 to expand a runway for the adjacent Cleveland Hopkins International Airport. In 2007 Glenn's 8- by 6-Foot Supersonic Wind Tunnel and the Abe Silverstein 10- by 10-Foot Supersonic Wind Tunnel were determined to be eligible for the National Register because of their contributions to the space shuttle program.[584]

As part of a centerwide Ohio Historic Inventory report in the mid-1990s, Gray & Pape, Inc., conducted a survey of the AWT in May 1996. Although they neglected to make a determination on its National Register eligibility, the report states, "The Altitude Wind Tunnel has been cited as historically the most important facility at [NASA Lewis Research Center]."[585] The AWT's

Image 278: The Icing Research Tunnel, which sat in the shadow of the AWT for over 60 years, was named an International Historic Mechanical Engineering Landmark in 1987 by ASME International. (NASA C–2007–02576)

significance had been noted 10 years before by Virginia Dawson, who was then writing her definitive history of the center, *Engines and Innovation* (Ref. 151). Dawson was asked to coauthor a proposal to nominate the Icing Research Tunnel for an ASME International national landmark. She declined, stating, "As far as I am concerned, it is the Altitude Wind Tunnel, not the Icing Research Tunnel that is historically significant."[586]

Glenn is now (2009) in the process of creating a historic district for its main campus area. The district was found to have significance because of its associations with the technical advancement of national aeronautics and space programs. In addition, the center's original buildings were constructed with matching blonde brick facades with rustication and banded windows that exemplify an early 20th century research facility. Buildings added later were constructed in the same style, giving the main campus area a unified appearance. The AWT's Shop and Office Building was among those original structures. Since the tunnel portion of the AWT was slated for demolition prior to the demarcation of the Glenn historic district, it was not identified as contributing to the district.[587]

Historical Mitigation

In the fall of 1989 NASA signed an agreement with the National Conference of State Historic Preservation Officers and the Advisory Council on Historic Preservation that applies to the maintenance of NASA's National Historic Landmarks.[588] The National Historic Preservation Act requires federal agencies to identify and protect historic properties; its implementing regulations mandate that formal consultation be completed whenever a federal agency proposes a program or project that may affect a historic property. The National Historic Preservation Act created the Advisory Council on Historic Preservation, a small federal agency located in Washington, DC, to oversee regulatory compliance; and it authorized the individual State Historic Preservation Officers to serve as the regulatory authority during the consultation. The federal agency with the historic facility, the State Historic Preservation Officers, and the Advisory Council on Historic Preservation must reach an agreement on an appropriate level of documentation, or mitigation, of a facility prior to any construction work.[589]

There are properties that are eligible for the National Register but are not formally nominated because of political, planning, or other reasons. The National Historic Preservation Act states that federal agencies must follow the same consultation process for eligible properties as they do with listed properties. Although there was no formal decision on the AWT's eligibility,

Glenn decided to treat the facility as eligible for the National Register. The center recognized that the planned demolition of the AWT was an adverse effect under the National Historic Preservation Act regulations. Joe Morris, the Glenn Historic Preservation Officer at the time, initiated consultation with Ohio State Historic Preservation Officer Lisa Adkins and the Advisory Council on Historic Preservation shortly after Glenn decided to demolish the AWT in 2004. The Advisory Council on Historic Preservation informed Morris that they did not need to be a consulting party and simply requested to be copied on the final agreement. Adkins visited Glenn on 22 August 2005 to meet with Morris and tour the AWT site.

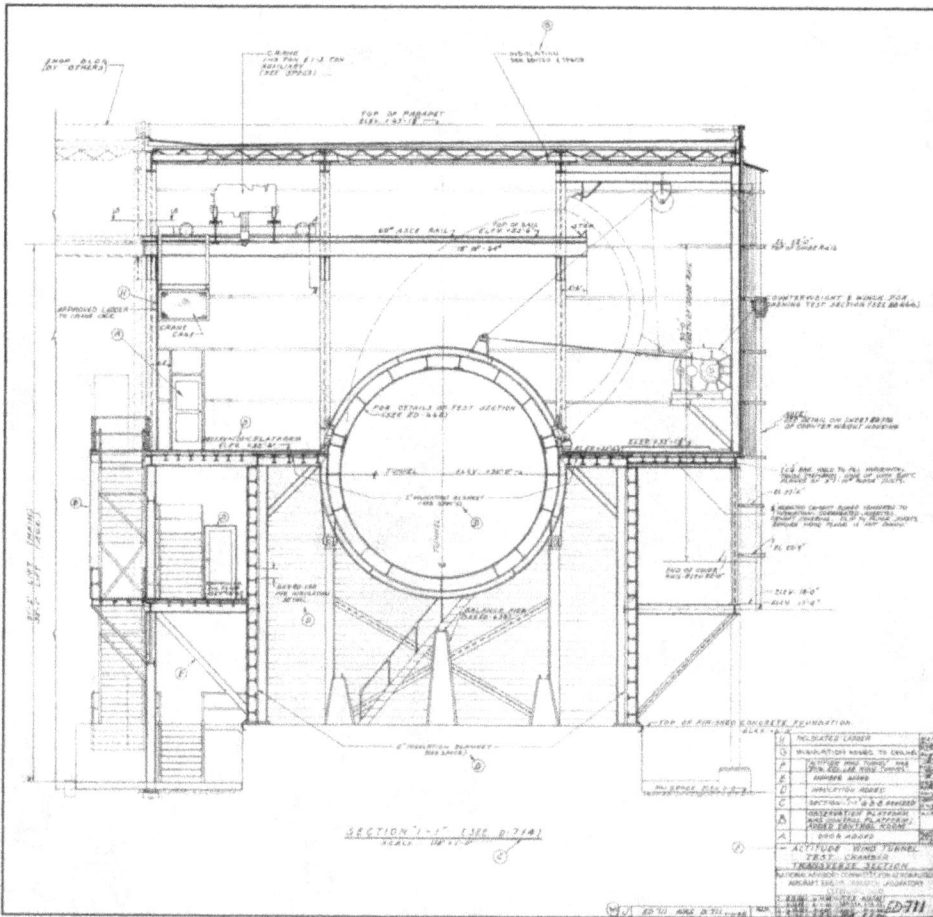

Image 279: The documentation of the AWT included gathering drawings (such as this cross section of the test section), conducting oral histories, researching archival collections, and studying technical reports. (1942) (NASA ED–71101)

Image 280: Left to right: Building Manager Wayne Condo and former NASA Glenn History Officer Kevin Coleman inside the SPC. Coleman was instrumental in brokering a deal with Glenn's Facilities Division to document the AWT before its demolition. (NASA C–2007–02567)

Image 281: On 22 August 2005, Ohio State Historic Preservation Officer Lisa Adkins met with Glenn officials to assess the mitigation required to document the AWT prior to its demolition. Left to right (background and foreground): Bob Arrighi, Kevin Coleman, Wayne Condo, Les Main, Joe Morris, Steve Gordon, Lisa Adkins, and Bob Houk (NASA C–2007–02564)

Image 282: NASA photographer Pete Tate and Bob Arrighi set up camera equipment inside SPC No. 1 in October 2005. The photographic survey of the facility included both still images and 360° panoramic images. (NASA C–2005–01653)

In May 2007 a Memorandum of Agreement was executed between Glenn and the Ohio State Historic Preservation Officer in accordance with the National Historic Preservation Act. The agreement outlines the mitigation measures that Glenn agreed to implement to offset the demolition of the historic property. NASA Glenn undertook a broad effort to both physically document the facility and compile the history of its construction, research, and contributions to the nation's aeronautics and space community. The Glenn History Office agreed to perform the mitigation work for Glenn's Facilities Division. Glenn's actions surpassed the mitigation measures that NASA normally takes when demolishing a historic property. The facility and its support buildings were extensively photographed and filmed prior to the AWT's removal. Documents, photographs, blueprints, films, and oral histories were gathered. A great deal of research was performed, as well, resulting in a website, an interactive computer disk (CD-ROM), and reports describing the tunnel and its history. These items serve as a permanent documentary record for the facility, lessons learned insight for internal NASA use, increased public awareness of NASA Glenn's contributions, educational resources, and a collected body of materials for future researchers.

Demolition

The AWT demolition focused on the actual tunnel and its support columns. The tunnel's test section within the former Shop and Office Building, the Exhauster Building, and the Refrigeration Building were not destroyed. The former Circulating Water Pump House and the Vacuum Pump House underneath the tunnel were included in the demolition. NASA Glenn also decided to fund several options not included in the NASA Headquarters package. These included removing the large generators and drive motor in the Exhauster Building and renovating the high-bay exterior.[590]

An Environmental Assessment was conducted to address land use, water resources, ambient noise, endangered species, hazardous materials, cultural resources, and other concerns. The assessment found that the only short-term concerns were air quality, noise, and land use due to the construction work. The only long-term concern was the removal of a historic property.[591]

Normal NASA demolition projects consist of three phases: relocation of the utilities, remediation of hazardous materials, and destruction of the structure. The relocation of utility lines and pipes began in early 2008. The AWT's original construction methods and certain aspects of its operational history posed potential environmental hazards, particularly regarding lead-based paint, asbestos ties between the shell layers, asbestos-containing transite siding on the high bay, and possible polychlorinated biphenyls (PCBs). The internal remediation of these hazards occurred in the fall of 2008. Most of the external remediation took place simultaneously with the demolition. In December workers on scaffolding began cutting away sections of the outer shell and safely removing the asbestos ties and insulation. Afterward the inner shell was segmented with torches and removed with a crane.[592]

The demolition work was largely completed in May 2009. The last major task, the removal of the drive motor and generators from the Exhauster Building, was completed the first week of June.

Photo Essay 9:
The Giant Comes Down

Image 283: Workers begin ripping off the outer shell. (NASA C-2008-045931)

Image 284: The insulation and steel mesh are removed from between the shells. (NASA C-2008-04591)

Image 285: A construction worker uses a torch to segment the inner shell. (NASA C-2008-04596)

Image 286. In December 2008 a gaping wound appeared in the AWT. (NASA C–2008–04475)

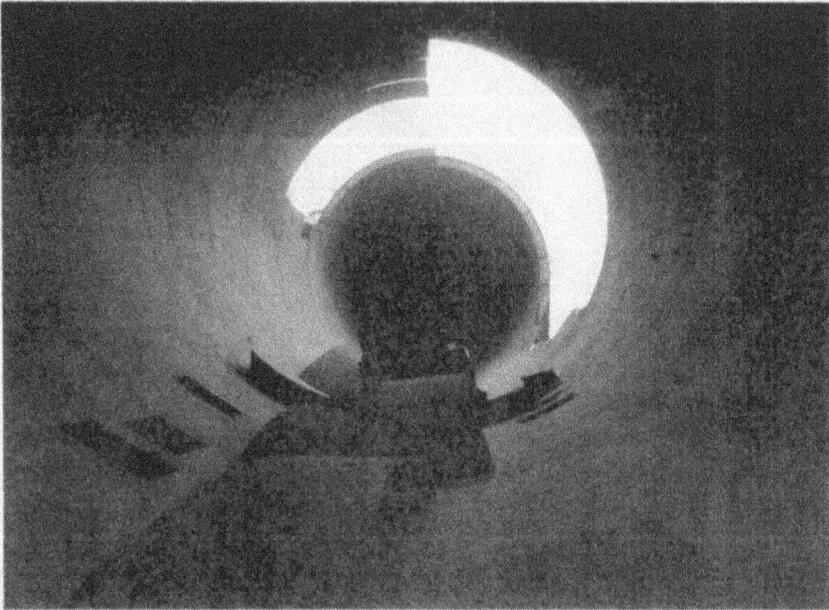

Image 287. Pieces of the shell were placed inside the tunnel for temporary storage. (NASA C–2008–04599)

Image 288: One of the last tasks (1 June 2009) was lifting the drive motor out. (NASA C–2009–01589)

Image 289: Workers clean up debris at the AWT demolition site. (NASA C–2009–00100)

Image 290. As shown in this 26 February 2009 photograph, the removal of the structure started in the middle of the tunnel and proceeded to the two ends with their larger corner rings. Pieces of the shell were flattened and laid on the muddy ground to support the large cranes. (NASA C–2009–00752)

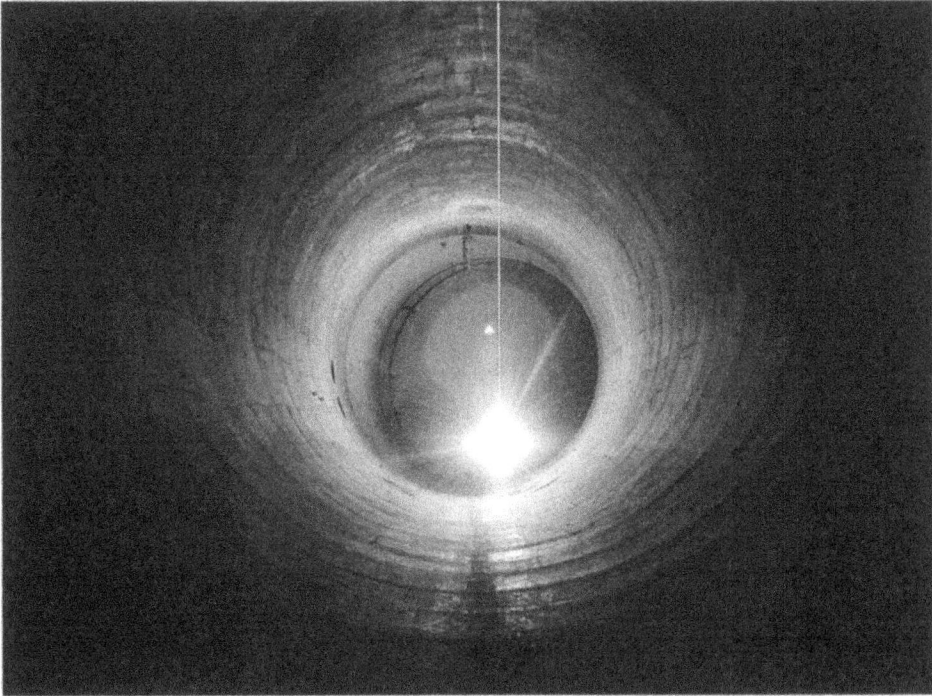

Image 291: A photographic survey of the AWT in April 2007 prior to its demolition. (NASA C–2007–00383)

Epitaph

The AWT/SPC facility was one of the most successful test facilities for both Lewis and the NACA and NASA agencies. Like most successful facilities, it adapted and modified itself over time to meet the changing needs of the aerospace community. The insightful design and the original investment in such a strong structure resulted in a facility robust enough to withstand these modifications.

The building itself, however, would be of minimal benefit if it were not for the excellence of the skilled staff and the farsighted leadership exemplified by Abe Silverstein. The research engineers worked closely with the facility operators and the mechanics on a daily basis to study and improve aircraft engines for years. With the advent of the space program, the staff adapted itself to the new technology and methods of handling it. Silverstein seemed one step ahead of the new technology, or at least ready to take it on the moment it became available. This was true of the original turbojets, the afterburner, supersonic tunnels, liquid hydrogen, the human space program, and Centaur. The AWT/SPC was involved in each of these projects.

The facility had not been significantly modified since the 1960s, and preventative maintenance ceased. NASA Glenn chose to invest in the restoration of the Space Power Facility and Space Propulsion Facility at Plum Brook Station for the new space initiative. Although there are still people who feel that the demolition is a loss, aeronautics research was ebbing and alternative uses of the facility were not feasible. The AWT/SPC is gone, but ironically, as a result of the demolition, its story has finally reached a larger audience.

Endnotes for Chapter 13

577. Philip S. Anton, et al., *Wind Tunnel and Propulsion Test Facilities: Supporting Analyses to an Assessment of NASA's Capabilities to Serve National Needs* (Santa Monica, CA: Rand Corporation, 2004).

578. Hearing Before the Subcommittee on Space and Aeronautics, "The Future of Aeronautics at NASA," 109th Congress, 1st Session, Serial No. 109–8, 16 March 2005.

579. Leslie Main, "Recordation of the Glenn Research Center Altitude Wind Tunnel Section 106 Check Sheets," NASA Glenn History Collection, Cleveland, OH.

580. Trudy Kortes, "AWT and PSL Environmental Assessment," Community Awareness Meeting at NASA Glenn Research Center, Cleveland, OH (27 April 2006).

581. NASA Glenn Facilities Division, "Project Requirements Document for Demolition of the Altitude Wind Tunnel Building 7 at Glenn Research Center. Project No. 9395," September 10, 2004, NASA Glenn History Collection, Cleveland, OH.

582. Angel L. Pagan, "NASA GRC Solicitation: Demolition of the Altitude Wind Tunnel," 19 July 2006.

583. Interview with Lesley Main and Bryan Coates, conducted by Bob Arrighi, 13 May 2008, NASA Glenn History Collection, Cleveland, OH.

584. Archaeological Consultants, Inc., "Survey and Evaluation of NASA-Owned Historic Facilities and Properties in the Context of the U.S. Space Shuttle Program: John H. Glenn Research Center at Lewis Field, Cleveland and Plum Brook Station, Sandusky, OH," November 2007.

585. Deborah McClane, Gray and Pape, Inc., *Ohio Historic Inventory No. CUY–4587–15* (Columbus, OH: Ohio Historical Society, May 1996).

586. Virginia P. Dawson to Dr. William Olsen, 21 March 1986, NASA Glenn History Collection, Cleveland, OH.

587. Main and Coates interview, conducted by Arrighi, 13 May 2008.

588. Programmatic Agreement Among the National Aeronautics and Space Administration, the National Conference of State Historic Preservation Officers, and the Advisory Council on Historic Preservation (1989).

589. Advisory Council on Historic Preservation, "National Historic Preservation Act of 1966, As Amended Through 2000," *http://www.achp.gov/NHPA.pdf* (accessed 8 October 2009).

590. NASA Glenn Facilities Division, "Project Requirements Document for Demolition of the Altitude Wind Tunnel Building 7 at Glenn Research Center. Project No. 9395," 10 September 2004, NASA Glenn History Collection, Cleveland, OH.

591. Kortes, "AWT and PSL Environmental Assessment."

592. Main and Coates interview, conducted by Arrighi, 13 May 2008.

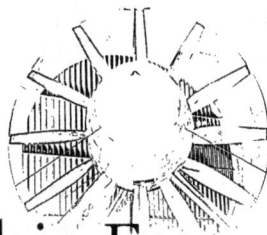

Bibliographic Essay

As the archivist at the NASA Glenn Research Center, I was both familiar with and had access to many materials in our History Collection that were used for this book. Documents such as historical correspondence and reports from our Director's Office, talks given at the National Advisory Committee for Aeronautics's (NACA's) triennial inspections, the complete run of our center newspaper, and oral history transcripts provided material for a significant portion of the manuscript.

In addition, my close association with Glenn's Imaging Technology Center provided me access to the center's extensive photograph and film collections. Though the original captions were often vague, the dates of photographs and ancillary information from the image often helped piece together the Altitude Wind Tunnel (AWT) story. Most of the information about the tests conducted in the AWT and Space Power Chambers (SPC) was gathered from the massive archive of NACA and NASA technical reports. I was also able to study scores of blueprints on the facility in all of its incarnations.

I conducted 13 interviews with eight retirees and one with two current employees. Bill Harrison, Harold Friedman, and Frank Holt were able to describe the lab's earliest years. Howard Wine and Bob Walker were familiar with the 1950s and 1960s testing. Larry Ross and Joe Nieberding were authoritative on the Centaur period, Les Main and Bryan Coates provided information regarding the demolition of the tunnel. NASA Ames historian, Glenn Bugos, talked to retiree Walter Vincenti about the design of the AWT for me.

Virginia Dawson and Mark Bowles have written extensively on the history of NASA Glenn. Their *Engines and Innovation* and *Taming Liquid Hydrogen* go into further detail on some of the topics covered in this book. The source materials used by Dawson and Bowles are archived in the Glenn History Collection. This material, particularly the oral history transcripts, has been invaluable in bringing this story to life.

Documents regarding the proposed reinstitution of the wind tunnel in the 1980s were found in the Glenn Records Management System. Les Main and Tina Norwood supplied much of the information regarding NASA historic properties, the demolition project, and the Historic Preservation Act. Several of these were documents filed by NASA Glenn with the Ohio Historic Preservation Office.

I gathered oral history transcripts from the NASA Johnson Space Center Oral History Project and the NASA Headquarters History Collection. I was able to locate additional interviews with Abe Silverstein and correspondence by George Lewis and Charles Lindbergh at NASA Headquarters. Copies of all materials used for this publication were added to the Glenn History Collection.

Numerous secondary resources were consulted for contextual information. The most important were Dawson's *Engines and Innovation*, General Electric's *Five Decades of Progress*, Margaret Engel's *Willis Haviland Carrier*, William Fleming's *History of North American Small Gas Turbines*, James Hansen's *Engineer in Charge*, James Grimwood's *This New Ocean*, Robert Dorr's *B-29 Superfortress*, and T. A. Heppenheimer's "The Jet Airplane Is Born." Other useful resources included NASA's National Space Science Data Center, Eugene Emme's Aeronautics and Astronautics timelines, and the National Museum of the Air Force.

Interview List

Bechtel, Frank, 19 July 2004 (Cleveland, OH)

Friedman, Harold, 2 Nov. 2005 (Beachwood, OH)

Harrison, Billy, 9 Aug. 2005, 16 Sept. 2005, 14 Oct. 2005, and 8 Aug. 2008 (North Olmsted, OH)

Holt, Frank, 9 Aug. 2005 and 9 Sept. 2005 (Cleveland, OH)

Main, Leslie and Bryan Coates, 13 May 2008 (Cleveland, OH)

Nieberding, Joseph, 16 Feb. 2008 (Cleveland, OH)

Ross, Larry, 1 Mar. 2007 (Cleveland, OH)

Walker, Robert, 2 Aug. 2005 (Cleveland, OH)

Wine, Howard, 4 Sept. 2005 (Cleveland, OH)

Website and CD-ROM

The documentation of the Altitude Wind Tunnel included the gathering of many historical materials. Many of these are available on the website: http://awt.grc.nasa.gov.

The home page has 360° panoramic photographs of the interior and exterior of the tunnel.

The Interactive History section launches a Flash multimedia piece that includes a detailed chronological history, facility layouts with photographs, a documentary video, a collection of related technical reports, and several hundred videos and images. (The Interactive History section is also available as a computer disk (CD-ROM). It can be obtained by sending a self-addressed, stamped envelope to the NASA History Division, Room CO72, NASA Headquarters, 300 E St. SW, Washington, DC 20546.)

The Facility section of the website describes the physical characteristics and operation of the wind tunnel and altitude chambers in detail. The pages include descriptions of the refrigeration system, control rooms, and other components.

National Aeronautics and Space Administration

ALTITUDE WIND TUNNEL
at NASA Glenn Research Center

An Interactive History

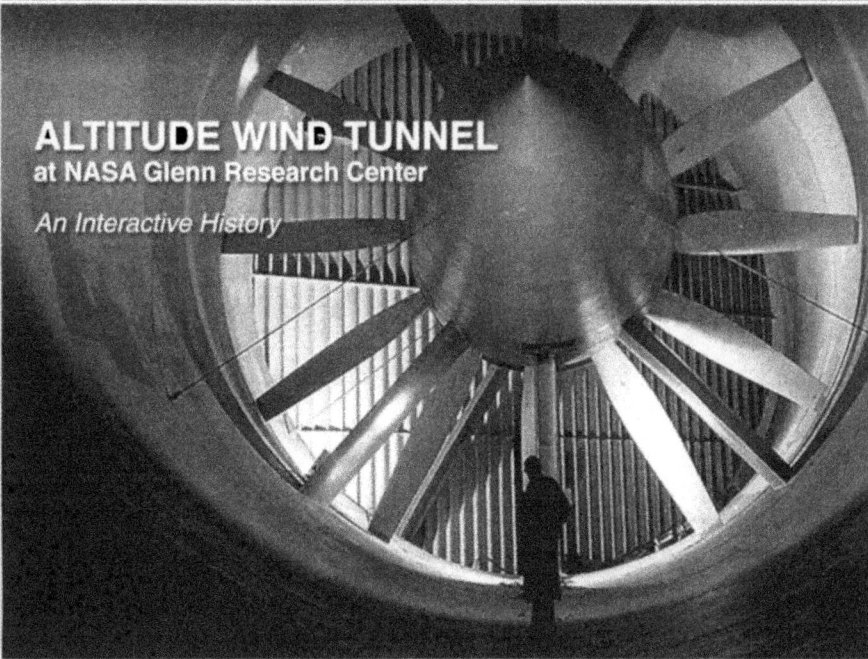

The Mitigation section describes the historical mitigation that was done prior to demolition of the facility and includes related documents and photographs.

The Research section has a two-part event timeline, a timeline of AWT/SPC tests, and Portable Document Format (PDF) files of 21 historical documents.

The Students section include short narrated animations that describe how the AWT and SPC worked, brief histories of wind tunnels and vacuum chambers, and glossaries.

The Gallery section includes over 1300 images, including captions and high-resolution versions.

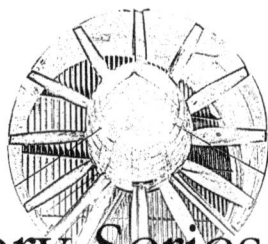

The NASA History Series

Reference Works, NASA SP-4000:

Grimwood, James M. *Project Mercury: A Chronology.* NASA SP-4001, 1963.

Grimwood, James M., and Barton C. Hacker, with Peter J. Vorzimmer. *Project Gemini Technology and Operations: A Chronology.* NASA SP-4002, 1969.

Link, Mae Mills. *Space Medicine in Project Mercury.* NASA SP-4003, 1965.

Astronautics and Aeronautics, 1963: Chronology of Science, Technology, and Policy. NASA SP-4004, 1964.

Astronautics and Aeronautics, 1964: Chronology of Science, Technology, and Policy. NASA SP-4005, 1965.

Astronautics and Aeronautics, 1965: Chronology of Science, Technology, and Policy. NASA SP-4006, 1966.

Astronautics and Aeronautics, 1966: Chronology of Science, Technology, and Policy. NASA SP-4007, 1967.

Astronautics and Aeronautics, 1967: Chronology of Science, Technology, and Policy. NASA SP-4008, 1968.

Ertel, Ivan D., and Mary Louise Morse. *The Apollo Spacecraft: A Chronology, Volume I, Through November 7, 1962.* NASA SP-4009, 1969.

Morse, Mary Louise, and Jean Kernahan Bays. *The Apollo Spacecraft: A Chronology, Volume II, November 8, 1962–September 30, 1964.* NASA SP-4009, 1973.

Brooks, Courtney G., and Ivan D. Ertel. *The Apollo Spacecraft: A Chronology, Volume III, October 1, 1964–January 20, 1966.* NASA SP-4009, 1973.

Ertel, Ivan D., and Roland W. Newkirk, with Courtney G. Brooks. *The Apollo Spacecraft: A Chronology, Volume IV, January 21, 1966–July 13, 1974*. NASA SP-4009, 1978.

Astronautics and Aeronautics, 1968: Chronology of Science, Technology, and Policy. NASA SP-4010, 1969.

Newkirk, Roland W., and Ivan D. Ertel, with Courtney G. Brooks. *Skylab: A Chronology*. NASA SP-4011, 1977.

Van Nimmen, Jane, and Leonard C. Bruno, with Robert L. Rosholt. *NASA Historical Data Book, Vol. I: NASA Resources, 1958–1968*. NASA SP-4012, 1976, rep. ed. 1988.

Ezell, Linda Neuman. *NASA Historical Data Book, Vol. II: Programs and Projects, 1958–1968*. NASA SP-4012, 1988.

Ezell, Linda Neuman. *NASA Historical Data Book, Vol. III: Programs and Projects, 1969–1978*. NASA SP-4012, 1988.

Gawdiak, Ihor, with Helen Fedor. *NASA Historical Data Book, Vol. IV: NASA Resources, 1969–1978*. NASA SP-4012, 1994.

Rumerman, Judy A. *NASA Historical Data Book, Vol. V: NASA Launch Systems, Space Transportation, Human Spaceflight, and Space Science, 1979–1988*. NASA SP-4012, 1999.

Rumerman, Judy A. *NASA Historical Data Book, Vol. VI: NASA Space Applications, Aeronautics and Space Research and Technology, Tracking and Data Acquisition/Support Operations, Commercial Programs, and Resources, 1979–1988*. NASA SP-4012, 1999.

Rumerman, Judy A. *NASA Historical Data Book, Vol. VIII: NASA Launch Systems, Space Transportation, Human Spaceflight, and Space Science, 1979–1988*. NASA SP-2009-4012.

There is no NASA SP-4013.

Astronautics and Aeronautics, 1969: Chronology of Science, Technology, and Policy. NASA SP-4014, 1970.

Astronautics and Aeronautics, 1970: Chronology of Science, Technology, and Policy. NASA SP-4015, 1972.

Astronautics and Aeronautics, 1971: Chronology of Science, Technology, and Policy. NASA SP-4016, 1972.

Astronautics and Aeronautics, 1972: Chronology of Science, Technology, and Policy. NASA SP-4017, 1974.

Astronautics and Aeronautics, 1973: Chronology of Science, Technology, and Policy. NASA SP-4018, 1975.

Astronautics and Aeronautics, 1974: Chronology of Science, Technology, and Policy. NASA SP-4019, 1977.

Astronautics and Aeronautics, 1975: Chronology of Science, Technology, and Policy. NASA SP-4020, 1979.

Astronautics and Aeronautics, 1976: Chronology of Science, Technology, and Policy. NASA SP-4021, 1984.

Astronautics and Aeronautics, 1977: Chronology of Science, Technology, and Policy. NASA SP-4022, 1986.

Astronautics and Aeronautics, 1978: Chronology of Science, Technology, and Policy. NASA SP-4023, 1986.

Astronautics and Aeronautics, 1979–1984: Chronology of Science, Technology, and Policy. NASA SP-4024, 1988.

Astronautics and Aeronautics, 1985: Chronology of Science, Technology, and Policy. NASA SP-4025, 1990.

Noordung, Hermann. *The Problem of Space Travel: The Rocket Motor.* Edited by Ernst Stuhlinger and J.D. Hunley, with Jennifer Garland. NASA SP-4026, 1995.

Astronautics and Aeronautics, 1986–1990: A Chronology. NASA SP-4027, 1997.

Astronautics and Aeronautics, 1991–1995: A Chronology. NASA SP-2000-4028, 2000.

Orloff, Richard W. *Apollo by the Numbers: A Statistical Reference.* NASA SP-2000-4029, 2000.

Lewis, Marieke and Swanson, Ryan. *Aeronautics and Astronautics: A Chronology, 1996-2000.* NASA SP-2009-4030, 2009.

Management Histories, NASA SP-4100:

Rosholt, Robert L. *An Administrative History of NASA, 1958–1963.* NASA SP-4101, 1966.

Levine, Arnold S. *Managing NASA in the Apollo Era*. NASA SP-4102, 1982.

Roland, Alex. *Model Research: The National Advisory Committee for Aeronautics, 1915–1958*. NASA SP-4103, 1985.

Fries, Sylvia D. *NASA Engineers and the Age of Apollo*. NASA SP-4104, 1992.

Glennan, T. Keith. *The Birth of NASA: The Diary of T. Keith Glennan*. Edited by J.D. Hunley. NASA SP-4105, 1993.

Seamans, Robert C. *Aiming at Targets: The Autobiography of Robert C. Seamans*. NASA SP-4106, 1996.

Garber, Stephen J., editor. *Looking Backward, Looking Forward: Forty Years of Human Spaceflight Symposium*. NASA SP-2002-4107, 2002.

Mallick, Donald L. with Peter W. Merlin. *The Smell of Kerosene: A Test Pilot's Odyssey*. NASA SP-4108, 2003.

Iliff, Kenneth W. and Curtis L. Peebles. *From Runway to Orbit: Reflections of a NASA Engineer*. NASA SP-2004-4109, 2004.

Chertok, Boris. *Rockets and People, Volume 1*. NASA SP-2005-4110, 2005.

Chertok, Boris. *Rockets and People: Creating a Rocket Industry, Volume II*. NASA SP-2006-4110, 2006.

Chertok, Boris. *Rockets and People: Hot Days of the Cold War, Volume III*. NASA SP–2009–4110, 2009.

Laufer, Alexander, Todd Post, and Edward Hoffman. *Shared Voyage: Learning and Unlearning from Remarkable Projects*. NASA SP-2005-4111, 2005.

Dawson, Virginia P., and Mark D. Bowles. *Realizing the Dream of Flight: Biographical Essays in Honor of the Centennial of Flight, 1903–2003*. NASA SP-2005-4112, 2005.

Mudgway, Douglas J. *William H. Pickering: America's Deep Space Pioneer*. NASA SP-2008-4113.

Project Histories, NASA SP-4200:

Swenson, Loyd S., Jr., James M. Grimwood, and Charles C. Alexander. *This New Ocean: A History of Project Mercury*. NASA SP-4201, 1966; reprinted 1999.

Green, Constance McLaughlin, and Milton Lomask. *Vanguard: A History*. NASA SP-4202, 1970; rep. ed. Smithsonian Institution Press, 1971.

Hacker, Barton C., and James M. Grimwood. *On Shoulders of Titans: A History of Project Gemini*. NASA SP-4203, 1977, reprinted 2002.

Benson, Charles D., and William Barnaby Faherty. *Moonport: A History of Apollo Launch Facilities and Operations*. NASA SP-4204, 1978.

Brooks, Courtney G., James M. Grimwood, and Loyd S. Swenson, Jr. *Chariots for Apollo: A History of Manned Lunar Spacecraft*. NASA SP-4205, 1979.

Bilstein, Roger E. *Stages to Saturn: A Technological History of the Apollo/ Saturn Launch Vehicles*. NASA SP-4206, 1980 and 1996.

There is no NASA SP-4207.

Compton, W. David, and Charles D. Benson. *Living and Working in Space: A History of Skylab*. NASA SP-4208, 1983.

Ezell, Edward Clinton, and Linda Neuman Ezell. *The Partnership: A History of the Apollo-Soyuz Test Project*. NASA SP-4209, 1978.

Hall, R. Cargill. *Lunar Impact: A History of Project Ranger*. NASA SP-4210, 1977.

Newell, Homer E. *Beyond the Atmosphere: Early Years of Space Science*. NASA SP-4211, 1980.

Ezell, Edward Clinton, and Linda Neuman Ezell. *On Mars: Exploration of the Red Planet, 1958–1978*. NASA SP-4212, 1984.

Pitts, John A. *The Human Factor: Biomedicine in the Manned Space Program to 1980*. NASA SP-4213, 1985.

Compton, W. David. *Where No Man Has Gone Before: A History of Apollo Lunar Exploration Missions*. NASA SP-4214, 1989.

Naugle, John E. *First Among Equals: The Selection of NASA Space Science Experiments*. NASA SP-4215, 1991.

Wallace, Lane E. *Airborne Trailblazer: Two Decades with NASA Langley's 737 Flying Laboratory*. NASA SP-4216, 1994.

Butrica, Andrew J., ed. *Beyond the Ionosphere: Fifty Years of Satellite Communications*. NASA SP-4217, 1997.

Butrica, Andrew J. *To See the Unseen: A History of Planetary Radar Astronomy*. NASA SP-4218, 1996.

Mack, Pamela E., ed. *From Engineering Science to Big Science: The NACA and NASA Collier Trophy Research Project Winners.* NASA SP-4219, 1998.

Reed, R. Dale. *Wingless Flight: The Lifting Body Story.* NASA SP-4220, 1998.

Heppenheimer, T. A. *The Space Shuttle Decision: NASA's Search for a Reusable Space Vehicle.* NASA SP-4221, 1999.

Hunley, J. D., ed. *Toward Mach 2: The Douglas D-558 Program.* NASA SP-4222, 1999.

Swanson, Glen E., ed. *"Before This Decade is Out . . ." Personal Reflections on the Apollo Program.* NASA SP-4223, 1999.

Tomayko, James E. *Computers Take Flight: A History of NASA's Pioneering Digital Fly-By-Wire Project.* NASA SP-4224, 2000.

Morgan, Clay. *Shuttle-Mir: The United States and Russia Share History's Highest Stage.* NASA SP-2001-4225.

Leary, William M. *We Freeze to Please: A History of NASA's Icing Research Tunnel and the Quest for Safety.* NASA SP-2002-4226, 2002.

Mudgway, Douglas J. *Uplink-Downlink: A History of the Deep Space Network, 1957–1997.* NASA SP-2001-4227.

Dawson, Virginia P., and Mark D. Bowles. *Taming Liquid Hydrogen: The Centaur Upper Stage Rocket, 1958–2002.* NASA SP-2004-4230.

Meltzer, Michael. *Mission to Jupiter: A History of the Galileo Project.* NASA SP-2007-4231.

Heppenheimer, T. A. *Facing the Heat Barrier: A History of Hypersonics.* NASA SP-2007-4232.

Tsiao, Sunny. *"Read You Loud and Clear!" The Story of NASA's Spaceflight Tracking and Data Network.* NASA SP-2007-4233.

Center Histories, NASA SP-4300:

Rosenthal, Alfred. *Venture into Space: Early Years of Goddard Space Flight Center.* NASA SP-4301, 1985.

Hartman, Edwin, P. *Adventures in Research: A History of Ames Research Center, 1940–1965.* NASA SP-4302, 1970.

Hallion, Richard P. *On the Frontier: Flight Research at Dryden, 1946–1981.* NASA SP-4303, 1984.

Muenger, Elizabeth A. *Searching the Horizon: A History of Ames Research Center, 1940–1976.* NASA SP-4304, 1985.

Hansen, James R. *Engineer in Charge: A History of the Langley Aeronautical Laboratory, 1917–1958.* NASA SP-4305, 1987.

Dawson, Virginia P. *Engines and Innovation: Lewis Laboratory and American Propulsion Technology.* NASA SP-4306, 1991.

Dethloff, Henry C. *"Suddenly Tomorrow Came . . .": A History of the Johnson Space Center, 1957–1990.* NASA SP-4307, 1993.

Hansen, James R. *Spaceflight Revolution: NASA Langley Research Center from Sputnik to Apollo.* NASA SP-4308, 1995.

Wallace, Lane E. *Flights of Discovery: An Illustrated History of the Dryden Flight Research Center.* NASA SP-4309, 1996.

Herring, Mack R. *Way Station to Space: A History of the John C. Stennis Space Center.* NASA SP-4310, 1997.

Wallace, Harold D., Jr. *Wallops Station and the Creation of an American Space Program.* NASA SP-4311, 1997.

Wallace, Lane E. *Dreams, Hopes, Realities. NASA's Goddard Space Flight Center: The First Forty Years.* NASA SP-4312, 1999.

Dunar, Andrew J., and Stephen P. Waring. *Power to Explore: A History of Marshall Space Flight Center, 1960–1990.* NASA SP-4313, 1999.

Bugos, Glenn E. *Atmosphere of Freedom: Sixty Years at the NASA Ames Research Center.* NASA SP-2000-4314, 2000.

There is no NASA SP-4315.

Schultz, James. *Crafting Flight: Aircraft Pioneers and the Contributions of the Men and Women of NASA Langley Research Center.* NASA SP-2003-4316, 2003.

Bowles, Mark D. *Science in Flux: NASA's Nuclear Program at Plum Brook Station, 1955–2005.* NASA SP-2006-4317.

Wallace, Lane E. *Flights of Discovery: An Illustrated History of the Dryden Flight Research Center.* NASA SP-4318, 2007. Revised version of SP-4309.

General Histories, NASA SP-4400:

Corliss, William R. *NASA Sounding Rockets, 1958–1968: A Historical Summary.* NASA SP-4401, 1971.

Wells, Helen T., Susan H. Whiteley, and Carrie Karegeannes. *Origins of NASA Names.* NASA SP-4402, 1976.

Anderson, Frank W., Jr. *Orders of Magnitude: A History of NACA and NASA, 1915–1980.* NASA SP-4403, 1981.

Sloop, John L. *Liquid Hydrogen as a Propulsion Fuel, 1945–1959.* NASA SP-4404, 1978.

Roland, Alex. *A Spacefaring People: Perspectives on Early Spaceflight.* NASA SP-4405, 1985.

Bilstein, Roger E. *Orders of Magnitude: A History of the NACA and NASA, 1915–1990.* NASA SP-4406, 1989.

Logsdon, John M., ed., with Linda J. Lear, Jannelle Warren Findley, Ray A. Williamson, and Dwayne A. Day. *Exploring the Unknown: Selected Documents in the History of the U.S. Civil Space Program, Volume I, Organizing for Exploration.* NASA SP-4407, 1995.

Logsdon, John M., ed, with Dwayne A. Day, and Roger D. Launius. *Exploring the Unknown: Selected Documents in the History of the U.S. Civil Space Program, Volume II, External Relationships.* NASA SP-4407, 1996.

Logsdon, John M., ed., with Roger D. Launius, David H. Onkst, and Stephen J. Garber. *Exploring the Unknown: Selected Documents in the History of the U.S. Civil Space Program, Volume III, Using Space.* NASA SP-4407, 1998.

Logsdon, John M., ed., with Ray A. Williamson, Roger D. Launius, Russell J. Acker, Stephen J. Garber, and Jonathan L. Friedman. *Exploring the Unknown: Selected Documents in the History of the U.S. Civil Space Program, Volume IV, Accessing Space.* NASA SP-4407, 1999.

Logsdon, John M., ed., with Amy Paige Snyder, Roger D. Launius, Stephen J. Garber, and Regan Anne Newport. *Exploring the Unknown: Selected Documents in the History of the U.S. Civil Space Program, Volume V, Exploring the Cosmos.* NASA SP-4407, 2001.

Logsdon, John M., ed., with Stephen J. Garber, Roger D. Launius, and Ray A. Williamson. *Exploring the Unknown: Selected Documents in the History of the U.S. Civil Space Program, Volume VI: Space and Earth Science.* NASA SP-2004-4407, 2004.

Logsdon, John M., ed., with Roger D. Launius. *Exploring the Unknown: Selected Documents in the History of the U.S. Civil Space Program, Volume VII: Human Spaceflight: Projects Mercury, Gemini, and Apollo.* NASA SP-2008-4407, 2008.

Siddiqi, Asif A. *Challenge to Apollo: The Soviet Union and the Space Race, 1945–1974.* NASA SP-2000-4408, 2000.

Hansen, James R., ed. *The Wind and Beyond: Journey into the History of Aerodynamics in America, Volume 1, The Ascent of the Airplane.* NASA SP-2003-4409, 2003.

Hansen, James R., ed. *The Wind and Beyond: Journey into the History of Aerodynamics in America, Volume 2, Reinventing the Airplane.* NASA SP-2007-4409, 2007.

Hogan, Thor. *Mars Wars: The Rise and Fall of the Space Exploration Initiative.* NASA SP-2007-4410, 2007.

Monographs in Aerospace History (SP-4500 Series):

Launius, Roger D., and Aaron K. Gillette, compilers. *Toward a History of the Space Shuttle: An Annotated Bibliography.* Monograph in Aerospace History, No. 1, 1992.

Launius, Roger D., and J. D. Hunley, compilers. *An Annotated Bibliography of the Apollo Program.* Monograph in Aerospace History No. 2, 1994.

Launius, Roger D. *Apollo: A Retrospective Analysis.* Monograph in Aerospace History, No. 3, 1994.

Hansen, James R. *Enchanted Rendezvous: John C. Houbolt and the Genesis of the Lunar-Orbit Rendezvous Concept.* Monograph in Aerospace History, No. 4, 1995.

Gorn, Michael H. *Hugh L. Dryden's Career in Aviation and Space.* Monograph in Aerospace History, No. 5, 1996.

Powers, Sheryll Goecke. *Women in Flight Research at NASA Dryden Flight Research Center from 1946 to 1995.* Monograph in Aerospace History, No. 6, 1997.

Portree, David S. F., and Robert C. Trevino. *Walking to Olympus: An EVA Chronology.* Monograph in Aerospace History, No. 7, 1997.

Logsdon, John M., moderator. *Legislative Origins of the National Aeronautics and Space Act of 1958: Proceedings of an Oral History Workshop.* Monograph in Aerospace History, No. 8, 1998.

Rumerman, Judy A., compiler. *U.S. Human Spaceflight, A Record of Achievement 1961–1998.* Monograph in Aerospace History, No. 9, 1998.

Portree, David S. F. *NASA's Origins and the Dawn of the Space Age.* Monograph in Aerospace History, No. 10, 1998.

Logsdon, John M. *Together in Orbit: The Origins of International Cooperation in the Space Station.* Monograph in Aerospace History, No. 11, 1998.

Phillips, W. Hewitt. *Journey in Aeronautical Research: A Career at NASA Langley Research Center.* Monograph in Aerospace History, No. 12, 1998.

Braslow, Albert L. *A History of Suction-Type Laminar-Flow Control with Emphasis on Flight Research.* Monograph in Aerospace History, No. 13, 1999.

Logsdon, John M., moderator. *Managing the Moon Program: Lessons Learned From Apollo.* Monograph in Aerospace History, No. 14, 1999.

Perminov, V. G. *The Difficult Road to Mars: A Brief History of Mars Exploration in the Soviet Union.* Monograph in Aerospace History, No. 15, 1999.

Tucker, Tom. *Touchdown: The Development of Propulsion Controlled Aircraft at NASA Dryden.* Monograph in Aerospace History, No. 16, 1999.

Maisel, Martin, Demo J. Giulanetti, and Daniel C. Dugan. *The History of the XV-15 Tilt Rotor Research Aircraft: From Concept to Flight.* Monograph in Aerospace History, No. 17, 2000. NASA SP-2000-4517.

Jenkins, Dennis R. *Hypersonics Before the Shuttle: A Concise History of the X-15 Research Airplane.* Monograph in Aerospace History, No. 18, 2000. NASA SP-2000-4518.

Chambers, Joseph R. *Partners in Freedom: Contributions of the Langley Research Center to U.S. Military Aircraft of the 1990s.* Monograph in Aerospace History, No. 19, 2000. NASA SP-2000-4519.

Waltman, Gene L. *Black Magic and Gremlins: Analog Flight Simulations at NASA's Flight Research Center.* Monograph in Aerospace History, No. 20, 2000. NASA SP-2000-4520.

Portree, David S. F. *Humans to Mars: Fifty Years of Mission Planning, 1950–2000*. Monograph in Aerospace History, No. 21, 2001. NASA SP-2001-4521.

Thompson, Milton O., with J. D. Hunley. *Flight Research: Problems Encountered and What They Should Teach Us*. Monograph in Aerospace History, No. 22, 2001. NASA SP-2001-4522.

Tucker, Tom. *The Eclipse Project*. Monograph in Aerospace History, No. 23, 2001. NASA SP-2001-4523.

Siddiqi, Asif A. *Deep Space Chronicle: A Chronology of Deep Space and Planetary Probes 1958–2000*. Monograph in Aerospace History, No. 24, 2002. NASA SP-2002-4524.

Merlin, Peter W. *Mach 3+: NASA/USAF YF-12 Flight Research, 1969–1979*. Monograph in Aerospace History, No. 25, 2001. NASA SP-2001-4525.

Anderson, Seth B. *Memoirs of an Aeronautical Engineer: Flight Tests at Ames Research Center: 1940–1970*. Monograph in Aerospace History, No. 26, 2002. NASA SP-2002-4526.

Renstrom, Arthur G. *Wilbur and Orville Wright: A Bibliography Commemorating the One-Hundredth Anniversary of the First Powered Flight on December 17, 1903*. Monograph in Aerospace History, No. 27, 2002. NASA SP-2002-4527.

Chambers, Joseph R. *Concept to Reality: Contributions of the NASA Langley Research Center to U.S. Civil Aircraft of the 1990s*. Monograph in Aerospace History, No. 29, 2003. NASA SP-2003-4529.

Peebles, Curtis, editor. *The Spoken Word: Recollections of Dryden History, The Early Years*. Monograph in Aerospace History, No. 30, 2003. NASA SP-2003-4530.

Jenkins, Dennis R., Tony Landis, and Jay Miller. *American X-Vehicles: An Inventory—X-1 to X-50*. Monograph in Aerospace History, No. 31, 2003. NASA SP-2003-4531.

Renstrom, Arthur G. *Wilbur and Orville Wright: A Chronology Commemorating the One-Hundredth Anniversary of the First Powered Flight on December 17, 1903*. Monograph in Aerospace History, No. 32, 2003. NASA SP-2003-4532.

Bowles, Mark D., and Robert S. Arrighi. *NASA's Nuclear Frontier: The Plum Brook Research Reactor.* Monograph in Aerospace History, No. 33, 2004. NASA SP-2004-4533.

Wallace, Lane and Christian Gelzer. *Nose Up: High Angle-of-Attack and Thrust Vectoring Research at NASA Dryden, 1979–2001.* Monograph in Aerospace History No. 34, 2009. NASA SP–2009–4534.

Matranga, Gene J., C. Wayne Ottinger, Calvin R. Jarvis, and D. Christian Gelzer. *Unconventional, Contrary, and Ugly: The Lunar Landing Research Vehicle.* Monograph in Aerospace History, No. 35, 2006. NASA SP-2004-4535.

McCurdy, Howard E. *Low Cost Innovation in Spaceflight: The History of the Near Earth Asteroid Rendezvous (NEAR) Mission.* Monograph in Aerospace History, No. 36, 2005. NASA SP-2005-4536.

Seamans, Robert C., Jr. *Project Apollo: The Tough Decisions.* Monograph in Aerospace History, No. 37, 2005. NASA SP-2005-4537.

Lambright, W. Henry. *NASA and the Environment: The Case of Ozone Depletion.* Monograph in Aerospace History, No. 38, 2005. NASA SP-2005-4538.

Chambers, Joseph R. *Innovation in Flight: Research of the NASA Langley Research Center on Revolutionary Advanced Concepts for Aeronautics.* Monograph in Aerospace History, No. 39, 2005. NASA SP-2005-4539.

Phillips, W. Hewitt. *Journey Into Space Research: Continuation of a Career at NASA Langley Research Center.* Monograph in Aerospace History, No. 40, 2005. NASA SP-2005-4540.

Rumerman, Judy A., Chris Gamble, and Gabriel Okolski, compilers. *U.S. Human Spaceflight: A Record of Achievement, 1961–2006.* Monograph in Aerospace History No. 41, 2007. NASA SP-2007-4541.

Dick, Steven J.; Garber, Stephen J.; and Odom, Jane H. *Research in NASA History.* Monograph in Aerospace History No. 43, 2009. NASA SP-2009-4543.

Merlin, Peter W. *Ikhana: Unmanned Aircraft System Western States Fire Missions.* Monograph in Aerospace History No. 44. NASA SP-2009-4544.

Fisher, Steven J., ed., with Shamin A. Rahman. *Remembering the Giants: Apollo Rocket Propulsion Development.* Monograph in Aerospace History No. 45. NASA SP–2009–4545.

Electronic Media (SP-4600 Series)

Remembering Apollo 11: The 30th Anniversary Data Archive CD-ROM. NASA SP-4601, 1999.

Remembering Apollo 11: The 35th Anniversary Data Archive CD-ROM. NASA SP-2004-4601, 2004. This is an update of the 1999 edition.

The Mission Transcript Collection: U.S. Human Spaceflight Missions from Mercury Redstone 3 to Apollo 17. SP-2000-4602, 2001.

Shuttle-Mir: the United States and Russia Share History's Highest Stage. NASA SP-2001-4603, 2002.

U.S. Centennial of Flight Commission presents Born of Dreams ~ Inspired by Freedom. NASA SP-2004-4604, 2004.

Of Ashes and Atoms: A Documentary on the NASA Plum Brook Reactor Facility. NASA SP-2005-4605.

Taming Liquid Hydrogen: The Centaur Upper Stage Rocket Interactive CD-ROM. NASA SP-2004-4606, 2004.

Fueling Space Exploration: The History of NASA's Rocket Engine Test Facility DVD. NASA SP-2005-4607.

Altitude Wind Tunnel at NASA Glenn Research Center: An Interactive History CD-ROM. NASA SP-2008-4608.

Conference Proceedings (SP-4700 Series)

Dick, Steven J., and Keith Cowing, ed. *Risk and Exploration: Earth, Sea and the Stars.* NASA SP-2005-4701.

Dick, Steven J., and Roger D. Launius. *Critical Issues in the History of Spaceflight.* NASA SP-2006-4702.

Dick, Steven J., ed. *Remembering the Space Age: Proceedings of the 50th Anniversary Conference.* NASA SP-2008-4703.

Societal Impact (SP-4800 Series)

Dick, Steven J., and Roger D. Launius. *Societal Impact of Spaceflight.* NASA SP-2007-4801.

Dick, Steven J. ed., with Mark L. Lupisella. *Cosmos and Culture: Cultural Evolution in a Cosmic Context.* NASA SP–2009–4802.

List of Images

Preface

Chapter 1

Chapter 2

Chapter 3

Chapter 4

Chapter 6

Chapter 7

Chapter 9

Chapter 10

Chapter 11

Chapter 12

Chapter 13

Index

A

Bold page numbers refer to information in an image.

www.ingramcontent.com/pod-product-compliance
Lightning Source LLC
Chambersburg PA
CBHW082002190326
41458CB00010B/3043